Baiame, the creator Spirit Emu, left the earth after its creation to reside as a dark shape in the Milky Way. The emu is inextricably linked with the wide grasslands of Australia, the landscape managed by Aboriginals. The fate of the emu, people, and grain are locked in step because, for Aboriginal people, the economy and the spirit are inseparable. Europeans stare at the stars, but Aboriginal people also see the spaces in between where the Spirit Emu resides.

DARK EMU

Dark Emu won both the Book of the Year Award and the Indigenous Writer's Prize in the 2016 New South Wales Premier's Literary Awards.

Bruce Pascoe, who has been writing for many years, is currently working on two films for ABC TV and a novel. He lives at Gipsy Point, Victoria, and has a Bunurong, Tasmanian, and Yuin heritage.

Photograph courtesy Kim Batterham

Dark Emu

Aboriginal Australia and the birth of agriculture

BRUCE PASCOE

SCRIBE
Melbourne • London

Scribe Publications
18–20 Edward St, Brunswick, Victoria 3056, Australia
2 John St, Clerkenwell, London, WC1N 2ES, United Kingdom
3754 Pleasant Ave, Suite 100, Minneapolis, Minnesota 55409 USA

First published in Australia by Magabala Books Aboriginal Corporation 2014
Published by Scribe 2018
Reprinted 2019 (twice), 2020, 2021

Printed and bound in the UK by CPI Group (UK) Ltd, Croydon CR0 4YY

Scribe is committed to the sustainable use of natural resources
and the use of paper products made responsibly from those resources.

978 1 947534 08 7 (US edition)
978 1 911344 78 0 (UK edition)
978 1 925548 66 2 (e-book)

Catalogue records for this book are available from the British Library.

scribepublications.co.uk
scribepublications.com

To the Australians

Contents

Introduction

After my book on our colonial frontier battles, *Convincing Ground*, was published in Australia in 2007, I was inundated with more than 200 letters and emails — many of them from fourth-generation farmers and Aboriginal people. Farmers sent me their great grandparents' letters and documents about the frontier war, and Aboriginal people sent new information on many of those same battles.

I already had a pile of information collected from research conducted too late to make it into *Convincing Ground*, and, after following the leads from correspondents, I discovered much more.

I began to see a consistent thread running through the material: not only that the frontier war had been misrepresented in what we had been taught in school, but also that the economy and culture of Aboriginal and Torres

Strait Islander people had been grossly undervalued.

I knew that if I were to use all the new material in another book, I would have to begin from the sources upon which Australia's idea of history is based: the journals and diaries of explorers and colonists.

These journals revealed a much more complicated Aboriginal economy than the primitive hunter-gatherer lifestyle we had been told was the simple lot of Australia's First People. Hunter-gatherer societies forage and hunt for food, and do not employ agricultural methods or build permanent dwellings; they are nomadic. But as I read these early journals, I came across repeated references to people building dams and wells; planting, irrigating, and harvesting seed; preserving the surplus and storing it in houses, sheds, or secure vessels; and creating elaborate cemeteries and manipulating the landscape — none of which fitted the definition of a hunter-gatherer. Could it be that the accepted view of Indigenous Australians simply wandering from plant to plant, kangaroo to kangaroo, in hapless opportunism, was incorrect?

It is exciting to revisit the words of the first Europeans to 'witness' the pre-colonial Aboriginal economy. In *Dark Emu*, my aim is to give rise to the possibility of an alternative view of pre-colonial Aboriginal society. In reviewing the industry and ingenuity applied to food production over millennia, we have a chance to catch a glimpse of Australia as Aboriginals saw it.

Many readers of the explorers' journals see the hardships they endured, and are enthralled by their finds of grassy plains, bountiful rivers, and sites where great towns could be built; but by adjusting our perspective by only a few degrees, we see a vastly different world through the same window.

The first colonists had their minds wrought by ideas of race and destiny; by the rumours heard as children of the great British Empire. They were immersed in these stories as infants, and later while marching in to school to 'Men of Harlech', standing to attention for 'God Save the King', and poring breathlessly over the stories of Horatio Nelson, the Christian Crusaders, King Arthur, Oliver Cromwell, and, of course, Captain James Cook.

Europe was convinced that its superiority in science, economy, and religion directed its destiny. In particular, the British believed that their successes in industry accorded their colonial ambition a natural authority, and that it was their duty to spread their version of civilisation and the word of God to heathens. In return, they would capture the wealth of the colonised lands.

Charles Darwin's theory of evolution was still to come, but the basis of it, the gradual ascent from beast to civilised man, dominated the psychology of Europe at the time. The first British visitors sailed to Australia contemplating what they were about to find, and innate superiority was the prism through which their new world was seen.

When Darwin's theory was put forward, it gave comfort to those who believed it was their right and duty to occupy the 'empty' land. As anthropologist Tony Barta commented:

> The basis of that view was historical: it held that the advance of civilization was a triumphal progress, morally justifiable and probably inevitable. When Darwin lent his great gifts and influence to making the disappearance of peoples 'natural' as well as historical, his theory ... could serve as an ideological cover for policies abhorrent to his humanitarian and humanist principles. Darwin's fateful confusion of natural history and human history would be exploited fatally by others.[1]

Under the influence of these cultural certainties, how would it have been possible for the colonists not to believe that Englishmen were on the steepest ascent of human endeavour? How would it have been possible for them not to believe that the world was their entitlement, and their possession of it ordained by their God?

To understand how the Europeans' assumptions selectively filtered the information brought to them by the early explorers is to see how we came to have the history of the country we accept today. Linda Tuwahi Smith provides an analysis of imperialism, which reveals that it is more than an economic and military exercise; it's an act of ideology, the blatant confidence to see 'others' as tools for the will of the European.[2]

It is clear from the journals of the explorers that few were in Australia to marvel at a new civilisation; they were there to replace it. Most were simply describing a landscape from which settlers could profit. Few bothered with the evidence of the existing economy, because they knew it was about to be subsumed.

Skewed views and misconceptions

The following story serves as a good example of the power of these assumptions and the need for colonists to legitimise their presence in the colonial field.

The Beveridge family had prospered on the colonial plains around Melbourne to the degree that a district was named after them. Once their wealth was consolidated, they decided to send a son, Peter, and his friend, James Kirby, to an area of the Murray River that had never seen European occupation.

The young men drove 1,000 head of cattle from the outskirts of Melbourne to the Murray River in 1843. They came across some natives, and Beveridge wrote in his diary:

> [M]any of them had green boughs in their hands, and after 'yabber yabber' they began swinging the boughs over and round their heads, and shouting 'Cum-a-thunga,

> cum-a-thunga.' We of course did not know what their meaning was by these antics, but we guessed that by it they meant we were welcome to their land, and we made them understand that we were highly pleased at their antics and quite delighted at the words 'cum-a-thunga'. nWhen they saw we were so much pleased at their conduct, three or four of them jumped into the water, and swam across and gave us a lot more 'cum-a-thunga,' so much so that they almost made themselves hoarse with shouting 'cum-a-thunga'.[3]

You would have had to work hard to convince yourself, or the governor, that Aboriginal people were delighted to give away their land.

In subsequent days, the two young colonials observed substantial weirs built all through the river system, and speculated about who might have built them. As they were the first Europeans in the area, they conceded that they were probably built by the 'blacks'.

Later, they witnessed the people fishing with canoes, lines, and nets. The purpose of the weirs gradually became clear. They were made by damming the stream behind large earthen platforms into which channels were let, in order to direct fish as required. On one particular day, Kirby noticed a man by one of these weirs. He wrote:

> [A] black would sit near the opening and just behind him a tough stick about ten feet long was stuck in the

> ground with the thick end down. To the thin end of this rod was attached a line with a noose at the other end; a wooden peg was fixed under the water at the opening in the fence to which this noose was caught, and when the fish made a dart to go through the opening he was caught by the gills, his force undid the loop from the peg, and the spring of the stick threw the fish over the head of the black, who would then in a most lazy manner reach back his hand, undo the fish, and set the loop again around the peg.[4]

How did Kirby interpret this activity? After describing the operation in such detail, and appearing to approve of its efficiency, he wrote, 'I have often heard of the indolence of the blacks and soon came to the conclusion after watching a blackfellow catch fish in such a lazy way, that what I had heard was perfectly true.'[5]

Kirby's preconceptions of what he was going to find on this frontier are so powerful that he skews his detailed observations to that prejudice. The activity he witnessed was, in fact, a piece of ingenious engineering.

Peter Beveridge wrote a book about his experiences with Aboriginal people, in which he displayed all of his and Kirby's prejudices.[6] Despite the fact that his work is crucial to what we know of the Wati Wati clan, and that his list of words is one of the most significant, he can't disguise his contempt. He refers to the old women as hags, continually refers to the Wati Wati as savages, and appears to have completely ignored the moiety and totemic system of their society.

Modern histories of the area claim that Peter's brother, Andrew, was killed by the Wati Wati after a dispute about blacks killing Beveridge's sheep, but Kirby's description of the event offers a startling insight into the real motivation.

Heavily armed warriors advanced on the station and ignored all other Europeans until they found Andrew Beveridge, the man who they claimed had been violating women. He was isolated and speared, and his body symbolically daubed with ochre.[7]

The problems at the Beveridge property, Tyntynder, followed a very familiar colonial pattern: initial acceptance followed by increasing suspicion and anger as the Europeans refused to allow the people to make use of their ancestral lands.

Kirby relates incidents of the war with relish, but always cloaks the killings in euphemism:

> The blacks ran into the lake, but the shore shelved in so far that it was not deep enough for them to swim or dive, they thus became very good targets for us. A lot of these fellows never came near the hut again, nor did they attempt to kill a man or beast, no! they were very peaceable after this … Sir Robert [a Wati Wati man], for instance, *never killed anyone after this, he also may have died.*[8]

Kirby's emphatic words hint at a ghoulish glee.

His narrative continues: 'It was open war now. If they caught us unguarded they would kill us, and we in return would (if we caught them) *help ourselves*.'[9] The language Kirby uses may be euphemistic, but the meaning is unequivocal. Tyntynder was at war with the Wati Wati, despite the fact that at this stage of the settlement only one European had been killed by Aboriginals, and that was for molesting women.

When Kirby and Beveridge chose to interpret the Wati Wati shouts of 'cum-a-thunga' as an invitation to take their land, it set in train all the violence, bitterness, and hardship typical of the colonial frontier. It was a land contest, and neither side would withdraw from the battle.

In the dictionary he wrote in his retirement at French Island, Peter Beveridge does not give a definition of the first Aboriginal words he heard, but an examination of other studies, and discussions with linguists of the Wati Wati and the neighbouring Wemba Wemba language, reveal a phrase, 'cum.mar.ca.ta.ca', recorded by the Aboriginal Protector, George Augustus Robinson, one of the few who recorded language and cultural information. Its probable meaning is 'Get up and go away.' It's an exclamation given great force, as Beveridge admits, and it is improbable that it represents an invitation to take the land.

There is also a strong possibility that within the phrase heard by Beveridge is the word 'karmer', meaning a long reed spear, combined and added to the strongly intensive verb affix, 'ungga', and further combined with the plural first-person pronoun, we, 'angurr'. Thus, 'karmer ungga' translates as 'We will spear you.'

In any case, Beveridge chose not to include in his dictionary the first phrase of Wati Wati addressed to him. Perhaps he was not keen to remember it, having since learned the true meaning.

Kirby and Beveridge weren't just pulling their own legs; they were pulling ours in an effort to disguise the means by which they took possession of a land. Their determination to seize the land had blinded them to the use the Wati Wati were making of it. In denying the existence of the economy, they were denying the right of the people to their land, and fabricating the excuse that is at the heart of Australia's claim to legitimacy today.

Eric Rolls, in his epic *A Million Wild Acres*, described the desecration by sheep of the grasslands in the Hunter–Pillaga region. Rolls was a passionate man of the land who documented the misuse of soils and water by Australian farmers. He noticed that dispossession of Aboriginal people and destruction of their villages was followed by an equally rapid deterioration in the soil, the foundation of the pre-contact economy.

Farmers noticed the alarming drop in productivity over a mere handful of years as sheep ate out the croplands and compacted the light soils. 'In Australia thousands of years of grass and soil changed in a few years. The spongy soil grew hard, the run-off accelerated and different grasses dominated.'[10]

The fertility encouraged by careful husbandry of the soil was destroyed in just a few seasons. The lush yam pastures of Victoria disappeared as soon as sheep grazed upon them, as the dentition of sheep allowed them to eat growth right to the ground, destroying the basal leaves.

The English pastoralists weren't to know that the fertility they extolled on first entering the country was the result of careful management, and cultural myopia ensured that even as the nature of the country changed, they would never blame their own form of agriculture for that devastation.

At the height of its productivity, Australia supported large populations, and, even after plagues of introduced smallpox and warfare had devastated the Aboriginal population, 500 people attended the last ceremonies at Brewarrina in 1885. Similar reports of large gatherings were described in most parts of Australia around this time, despite the calamitous fall in population.

Colonial Australia sought to forget the advanced nature of the Aboriginal society and economy, and this amnesia was entrenched when settlers who arrived after the depopulation of whole districts found no structure more substantial than a windbreak, and no population that was not humiliated, debased, and diseased. This is understandable because, as is evidenced by the earlier first-hand reports, villages were burnt, the foundations stolen for other buildings, the occupants killed by warfare, murder, and disease, and the country usurped. It is no wonder that after 1860 most people saw no evidence of any prior complex civilisation.

Moreover, the perishable nature of materials used in Aboriginal storage devices ensured they would not be seen by archaeologists, and the ferocity of the war meant that such large stores of food could never be compiled again. The attacks by settlers on Aboriginals engaged in harvesting are much under-rated as one of the tools of war. Nutrition and morale plummeted as the croplands were mown down by sheep and cattle, and people were prevented from protecting and utilising their crops. No better device, short of murder, could ensure the weakening of the enemy.

The archaeologist Peter White, in his 'Agriculture: was Australia a bystander?', argues that de-population by disease and the arrival of sheep, which walked ahead of their shepherds, helped eliminate evidence of agriculture and its domesticates. This makes the evidence recorded by the earliest explorers and settlers so critical to our understanding of the pre-contact Aboriginal economy.

1
Agriculture

The use of the word 'agriculture' in relation to Australian Aboriginal people is not something many Australians would have heard. However, if we go back to the country's very first records of European occupation, we discover some extraordinary observations that provide a picture of what the Australian explorers and pioneers witnessed, and how it refutes the notion that Aboriginal people were only hunter-gatherers.

When Europeans began their classification of eras and the peoples of the world, they decided that five activities signified the development of agriculture: selection of seed, preparation of the soil, harvesting of the crop, storage of the surpluses, and erecting permanent housing for large populations.[1]

Rupert Gerritsen outlined the various theories on the

preconditions for incipient agriculture, but concluded that Australia may have gone well beyond the incipient stage.

'People farmed in 1788, but were not farmers,' Bill Gammage declared, and went on to say:

> These are not the same: one is an activity, the other a lifestyle. An estate may include a farm, but this does not make an estate manager a farmer ... In 1788 similarly, people never depended on farming. Mobility was much more important. It let people tend plants and animals in regions impossible for farmers today, and manage Australia more sustainably than their dispossessors. It was the critical difference between them and farmers ... Europeans think farming explains the difference between them and Aborigines. There must be a way of exploring those differences and their momentous consequences.[2]

We need to know more. We need more people to know it, so let us have another look at what the first Europeans saw.

Imagine you are riding beside the explorer and surveyor Major Thomas Mitchell (1792–1855). He's an educated and sensitive man, and great company, if a little eccentric. He's a great bushman, and a poet and painter, but also a hot head. Under some circumstances, he is obstinate and difficult, and is credited with fighting the last duel in

Australia, although he only succeeded in shooting a hole in his opponent's hat.

As he crosses the Australian frontier, he describes what he sees: '... [T]he grass is pulled ... and piled in hayricks, so that the aspect of the desert was softened into the agreeable semblance of a hay-field ... we found the ricks or hay-cocks extending for miles.'[3]

And later:

> [T]he seed is made by the natives into a kind of paste or bread. Dry heaps of this grass, that has been pulled expressly for this purpose of gathering the seed, lay along our path for many miles. I counted nine miles along the river, in which we rode through this grass only, reaching to our saddle-girths, and the same grass seemed to grow back from the river, at least as far as the eye could reach through a very open forest.[4]

Charles Sturt, on his journeys into South Australia and Queensland, also noticed the system of stacking grain into haycocks ready for threshing. Just as importantly, he commented on the frequency with which he encountered large, solidly built houses.

Mitchell also recorded his astonishment at the size of the villages. He noticed:

> [S]ome huts... being large, circular; and made of straight rods meeting at an upright pole in the centre;

> the outside had first been covered with bark and grass, and the entirety coated over with clay. The fire appeared to have been made nearly in the centre; and a hole at the top had been left as a chimney.[5]

He counts the houses, and estimates a population of over one thousand. He's disappointed that nobody's home; it's obvious they have only just left, and the evidence is everywhere that they have used the place for a very long time.

One of Mitchell's party comments that the buildings were 'of very large dimensions, one capable of containing at least 40 persons and of very superior construction'.[6]

If you had been with explorer George Grey in Western Australia in 1839, you might have wondered about the wisdom of your decision. Grey had no bush experience other than schoolboy idolatry of British explorers, and his Kimberley adventure was a disaster. The whale boats, overloaded and ill designed for the assignment, were wrecked on the beach at Gantheaume Bay, and the party had to walk the remaining distance to Perth.

Thankfully, Grey was a prolific diarist and, despite his predicament, he recorded all that he saw. He was surprised to find a village on the Gascoyne River, where the houses were 'built of large-sized logs, much higher, and altogether of a very superior description to those made by the natives of the south-western coast'.[7]

He was even more surprised to find land that appeared to have been cultivated. He wrote:

> [Fell] in with the native path we quitted yesterday; but now became quite wide, well beaten and differing altogether by its permanent character, from any I had seen in the southern part of this continent ... And as we wound along the native path my wonder augmented; the path increased in breadth and its beaten appearance, whilst along the side we found frequent wells, some of which were ten and twelve feet [3–4 metres] deep, and were altogether executed in a superior manner. We now crossed the dry bed of a stream, and from that emerged upon a tract of light fertile soil quite overrun with warran plants [the yam plant, *Dioscorea hastifolia*], the root of which is a favourite article of food with the natives. This was the first time we had seen this plant on our journey and now for three and a half consecutive miles [5.6 kilometres] traversed a piece of land, literally perforated with holes the natives made to dig this root; indeed we could with difficulty walk across it on that account whilst the tract extended east and west as far as we could see. It is now evident that we had entered the most thickly populated district of Australia that I had yet observed, and ... more had been done to secure provision from the ground by hard manual labour than I could believe it in the power of uncivilized man to accomplish. After crossing a low limestone range we

came upon another equally fertile warran ground … and (next day) passed two native villages, or as the men termed them, towns — the huts of which they composed differed from those in the southern districts, in being built, and very nicely plastered over the outside with clay, and clods of turf, so that although now uninhabited they were evidently intended for fixed places of residence.[8]

When John Batman, one of the founders of Melbourne and the colony of Victoria, left one of his men, Andrew Todd, to guard the stores at the first landing at Indented Head, Victoria, in June 1835, Todd whiled away the time with the local Wathaurong people, talking to them and sketching.

Yam diggers at Indented Head, Victoria, 1835
Yams were a staple of the First People's diet.

(J.H. Wedge)

One of these sketches shows a line of women digging for yam daisy, or murnong [*Microseris lanceolata*] tubers — a little sweet potato that was a staple vegetable of the Wathaurong. The area the women were working is perfectly clear, because they have made it so in order to most efficiently harvest their crop.

A handful of yams, and three-generations yam

(Vicky Shukuroglou)

In 1841, the Chief Aboriginal Protector of the Port Phillip District (1839–49), George Augustus Robinson, recorded:

> [T]he native women were spread out over the plain as far as the eye could see, collecting Murnong, or in this language pannin, a privilege they would not be permitted except under my protection. I inspected their bags and baskets on return and each had a load as much as she could carry.[9]

When Mitchell arrived at the Victorian Grampians in 1836, he saw 'a vast extent of open downs ... quite yellow with Murnong', and 'natives spread over the field, digging for roots'.[10] Captain John Hunter, captain on the First Fleet, reported in 1788 that the people around Sydney were dependent on their yam gardens.[11] 'The natives here, appear to live chiefly on the roots which they dig from the ground; for these low banks appear to have been ploughed up, as if a vast herd of swine had been living on them.'

In Sunbury, Victoria, in 1836, settlers, including Isaac Batey and Edward Page, observed that people had worked their gardens so well and for so long that large earthen mounds had been created during the process — but so little consideration was given to this land management that, only a few years later, Europeans couldn't say who or what had created these prominent terraces.

This last observation is evidence of a deliberate farming technique, one which any modern farmer would recognise as good soil management. The fact that explorers and settlers report seeing such activity in so many different parts of the country is an indication that it wasn't an isolated technique. Cultivation was a feature of Aboriginal land use.[12]

Charles Sievwright, the Assistant Protector of Aborigines of the Port Phillip District (1839–42) before it became the colony of Victoria, decided to introduce the European theory of farming to the Aboriginals assembled at his Lake Keilambete Protectorate. They took one look at his English ploughing technique, and immediately hoed the soil across the slope of the land and broke down all the larger clods. They'd been farming this land for thousands of years, and weren't about to allow erosion to ruin the land.

Similarly, Robinson, when entering the Mumbuller Valley near Pambula, New South Wales, was informed by a local elder, Yow.e.ge, that all the land thereabouts was his farm. The Yuin man was aware of the word that Europeans used for their food-production sites, and this comment indicates he was trying to impress on Robinson that his people were also cultivators.

Colonist Isaac Batey, when commenting on the disappearance of the yam daisy, remembers the women harvesting and washing the tubers in vast quantities. However, soon after his arrival in 1846 he notes:

> Where once abundant they have become quite extinct for the district where the writer was raised in this 1909 might be searched without discovering a solitary example ... Elsewhere it has been intimated that our domestic animals had eaten them out, yet there was another factor of destruction in the soil becoming hardened with the continuous tramping of sheep cattle or horses. In proof of that Mr Edward Page said 'when we first came here I started a vegetable garden, the soil dug like ashes.' It has to be added it was a spot free of

> timber or scrub of any description, the soil a reddish loam of great depth.[13]

Dr Beth Gott, a renowned ethnobotanist from the School of Biological Sciences at Monash University, has established a garden at the university with examples of plants eaten and used by Aboriginals before colonisation. In 'Ecology of Root Use by the Aborigines of Southern Australia', Gott explains that the effect of the systematic and repetitive tilling process aerated the soil, loosened it for seed germination and root penetration, and incorporated ash and compost material with the plants. She said that it 'bore sufficient resemblance to agriculture/horticulture to be regarded as a sort of natural gardening'.[14]

Archaeologist and Emeritus Professor David Frankel quotes the early observations of Batey:

> [T]he soil (on a sloping ridge) is rich in basaltic clay, evidently well fitted for the production of myrnongs [murnong, *Microseris lanceolata*]. On the spot are numerous mounds with short spaces between each, and as all these are at right angles to the ridge's slope it is conclusive evidence that they were the work of human hands extending over a long series of years. This uprooting of the soil, to apply the best term, was accidental gardening, still it is reasonable to assume that the aboriginals were quite aware of the fact that

> turning the earth over in search of yams, instead of diminishing that form of food supply, would have a tendency to increase it. On arriving in 1846 and thereafter myrnong digging was unknown to us, for the all sufficient reason that livestock seemingly had eaten out that form of vegetation.[15]

This is a description of terracing. So pronounced were the features that Batey was convinced they would endure for one hundred years.

The unusual quality and friability of the soil was reported by many colonists in the first years of settlement. The kangaroo grass in the Colac region of Western Victoria was so high it concealed the flocks of the first settler, G.T. Lloyd. Orchids, lilies, and mosses flourished among the grain crop, and: 'The ground had been so protected by mosses and lichens so thick that it was difficult to ride across the country at any pace exceeding the "farmers" jog trot.'[16] Lloyd says his horses sank to the fetlock into the soil as if it were sponge. 'With the onslaught of the sharp little hooves and teeth of herbivore sheep, goats, pigs and cattle driven in by the settlers, the ground covers were destroyed and the dews ceased.'[17] Once the soil hardened, rains ran off the compacted surfaces, and rivers flooded higher than the Aboriginals had ever seen. This created a new management problem for the soils of this district and others.

The persistent frequency of such colonial reports

inspired Gott to conduct her own experiment.[18] The Nodding Greenhood (*Pterostylis nutans*) was another significant tuberous food source for Aboriginal people, and the harvesting would have continually disturbed the soil as well as incorporating ash and compost below the surface. Gott found that after harvesting the greenhoods, 75 per cent of the pre-harvest density was restored within fourteen months. Harvesting on a cyclical mosaic over two to three years would see no diminution of supply, but would instead fertilise and enhance the crop.

These management practices created anomalous vegetation distributions. As Bill Gammage explained in *The Biggest Estate on Earth*, European settlers were surprised to find that the best Australian soils were virtually devoid of trees.

Aboriginal farmers had used fire to clear areas of land, which they were careful to separate with belts of timber. Like our contemporary farmers, Aboriginals left the forest on poorer soils and cleared the best soils so they could create pastures and croplands. Gammage quotes the memory of an early settler:

> With the exception of alluvial land, good timber is very seldom found on good land. The fertile plains in the interior are wholly destitute of it ... *[around Sydney]* the best forest lands are invariably thinnest of trees; and in general it will be found that the best lands are least encumbered with timber.[19]

There is still much to be learnt about Australian soils and how they were managed, but the explorers' journals suggest that colonial settlers ignored the Aboriginal method, and that contemporary Australians still suffer from the result.

The Aboriginal methods of land management were not just practical, but aesthetically pleasing. Mitchell noticed the beauty of the country, but considered it an accident: 'We crossed a beautiful plain; covered with shining verdure, and ornamented with trees, which, although "dropt in nature's careless haste", gave the country the appearance of an extensive park.'[20]

In a more pragmatic passage, Mitchell reported from the Belyando River, Central Queensland:

> We crossed some patches of dry swamp where the clods had been extensively turned up by the natives ... These clods were so very large and hard that we were obliged to throw them aside, and clear the way for our carts to pass. The whole resembled ground broken with the hoe ... There might be about two acres in the patch we crossed and we perceived at a distance other portions of the ground in a similar state.[21]

Near the Hunter River, in New South Wales, Mitchell noticed the peculiar furrowed appearance of the land, and pondered the cause of this feature.[22] In *A Million Wild Acres*, Rolls noted that settlers and surveyors of the district commented that, 'The hills have a look of a park and Grounds laid out.' Others saw what they considered 'ploughed land.'[23]

William Howitt reported on an intriguing land feature: 'The heaps or mounds are placed almost as regularly as the squares of a chess-board. Yet not exactly, for they are never in rows so as to allow you to pass between them. There is a heap and a hollow on every side of you.'[24] Howitt doesn't explain what they are, and attributes some natural phenomena at work. In his opinion, Aboriginal agency would be impossible; however, in the light of other observations of extensive cultivation, perhaps these were yet another example of a tilling technique.

Today the yam has virtually disappeared from the land, but a large field of this tuber has been discovered along the Bundian Way, near Delegate in New South Wales. By chance, sheep have never been grazed there, and superphosphate has never been distributed, so these unusual conditions allow us to study the yam in conditions similar to those in which traditional Aboriginals would have cultivated the tuber. The crop was so important to the Dhurga of New South Wales that they referred to themselves by their name for it. Another extensive field has been discovered near Port Campbell in Victoria. So, as we begin to look, we find.

The yam was a crucial plant in the economy of pre-colonial Aboriginal Australia, but few have examined this productive tuber. Surely we can no longer ignore such a valuable plant or the commercial opportunities it offers.

It is heartening, therefore, that Aboriginal communities in East Gippsland and the New South Wales south coast, encouraged by Beth Gott's research, are conducting experiments in yam cultivation. Seven plots on different soils and under different management regimes have been established, and scientific testing of the results is underway.

Left: The seed head of the yam daisy. Right: Drawing of yam and tuber.
(Beth Gott) (John Conran)

Grains

Even though it was convenient for many European settlers not to acknowledge the evidence of an Aboriginal agricultural economy, there were some who recorded their impressions, and of those, some speculated on what circumstances had produced this 'gentleman's park'. Records of the land's cultivated appearance are common in the early records, and are spread across the continent. These records were so persistent in their description of grain harvests from all parts of the country that Norman Tindale was able to plot Indigenous grain areas from which the Rural Industries Research and Development Corporation was able to construct the map reproduced below.

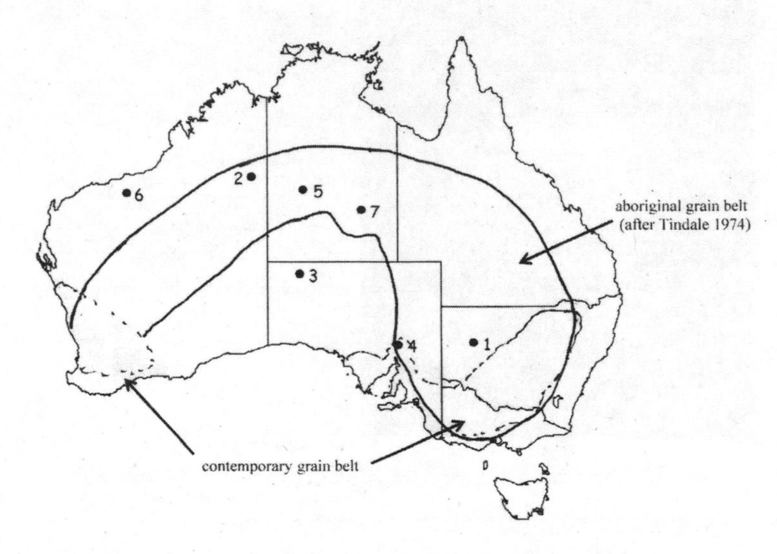

Aboriginal grain belt

The Aboriginal grain harvest map is based on the research compiled by Norman Tindale 1974, and shows the extent of the harvest compared to the current Australian grain belt:*

**1. Allen, 1974; 2. Cane, 1989; 3. Cleland and Johnston, 1936; 4. Cleland and Johnston 1939a; 5. Cleland andJohnston 1939b; 6. Maggiore, 1985; 7. O'Connell 1983.*

Areas beyond the high-rainfall zones of the coastal regions favoured grain as the staple crop, whereas in wetter areas yam production took over.

Tindale discovered that the people who harvested grain in these areas saw their methods as so central to their identity that they referred to themselves as 'grass people' by using the word 'panara', or similar forms of that word.

R.G. Kimber, a contemporary researcher and ethnographer, compiled an enormous body of evidence from people who observed Central Australian Aboriginals engaged in seed propagation, irrigation, harvest, storage, and the trade of seed across the region. One of Kimber's informants was the bush worker and cameleer Walter Smith.

Smith, who was proud of both his Welsh and Arabana Aboriginal heritage, told Kimber how seed was broadcast by hand, covered lightly with soil, and irrigated:

> They chuck a bit there [at a favourable locality]. Not much, you know, wouldn't be a handful. [They] chuck a little bit, spread it [broadcasting fashion] you see — one seed there, one seed there ... [of] course they chuck a little bit of dirt on, not too much though. And soon as first rain comes ... it will grow then.[25]

Smith is describing the method of taking seed into other areas where it didn't occur naturally, and trading it for other goods or giving it as simple gifts of reciprocity.

The idea of agriculture was so well advanced that seed was traded as a cultural item.[26] Several explorers and commentators witnessed grain in small sealed parcels being traded to distant relatives. This selection and trade of seed over such a wide area and such a long period gradually changed the morphology of the grains and other Aboriginal food sources — and produced the qualities that agriculturalists recognise as being the result of domestication.

The science of baking developed alongside the seed harvests. Richard Fullagar, at the Australian Museum, and Judith Field, at the University of New South Wales, found grindstones at Cuddie Springs, near Walgett, in western New South Wales, which had been used to grind seeds more than 30,000 years ago. This makes these people the world's oldest bakers by almost 15,000 years, as the Egyptians, the next earliest, didn't bake until 17,000 BC.[27] Other peoples ground tubers to extract starch, but it seems that Aboriginal people were the first to discover the alchemy of baking bread from the flour of grass seeds.

And this baking was not a one-off occurrence. Archaeologists found a 25,000-year-old grindstone at distant Kakadu in the Northern Territory: the bakers of antiquity. Why don't our hearts fill with wonder and pride?[28]

Alice Duncan-Kemp, who grew up with Aboriginal people on her father's station, Mooraberry, near Bidourie in Queensland around 1910, described the Katoora ceremony where:

> From their woven dilly bags the gins sprinkled seed food over the ground ... Katoora or barley grass seed lay in little hillocks, already swelling and creeping to repeated applications of water which the gins poured on them to make 'wunjee aal the same walkabout (grass to grow)'.[29]

The explorer Hamilton Hume, in frank conversation with Robinson, said that he had been on the exploration parties of Captain Charles Sturt (1795–1869), and that, 'on the Darling the Natives gather grain from the wild oats (a round grain) and grind it between two stones and make a paste and eat it, the same is done by the Natives to the northward.'[30]

Typical of the jealousy prevalent among the explorers, Sturt said that, 'Mitchell says nothing of this, indeed knows but little of their customs.'[31] It's true, however, that when Mitchell observed haycocks extending for miles over a stubble paddock where every stalk had been cut, he wondered if 'the heaps of grass had been pulled here, for some purpose connected with the allurement of agriculture'.[32]

When in 1845 Sturt first saw the harvested grass *Panicum laevinode* near Lake Torrens, he found it spread out to dry and ripen on the sloping banks of a stream.

It's significant that the people were engaged in a harvest while Sturt and his party were struggling to stay alive. 'The heat during the day had been terrific ... we were unable to keep our feet in the stirrups, and the horses perspired greatly.'[33] One of the party, Poole, suffered so badly from the heat and scurvy that his muscles stiffened and the roof of his mouth fell away. He was sent back to the base, but died on the way.

In this land of extreme heat and aridity, the Aboriginal inhabitants had built comfortable houses and produced grain surplus to their immediate requirements. This is an important social and economic achievement — surplus food production is one of the acknowledged characteristics of sedentary agriculture.

Even further to the north, Sturt saw 'grassy plains spreading out like a boundless stubble field, the grass being of the kind from which the natives collect seed for subsistence at this season of the year ... large heaps that had been thrashed out by the natives were piled up like haycocks.'[34] A boundless field of stubble? Haycocks? Sturt was observing a major harvest that must have provided a great surplus for the large number of people known to inhabit the region. Regardless of the evidence before his eyes, Sturt resorts automatically to the word 'subsistence'.

On a later expedition, one of Sturt's party, Brock, recorded his impression of land near Evelyn Creek: '[H]ere it is quite like a harvest field ... In every hollow we found the remains of the natives' labour in the shape of the straw from which they had beaten out the seed.'[35]

The party remarked on the prodigious quantities of grain harvested. This was the same grass Mitchell had seen in other areas, *Panicum decompositum*, commonly called barley grass or native millet, and known to Aboriginals as cooly or parpar. In fact, one of the areas Sturt visited in 1845 was called Parpir, and his journal records that they had been riding through vast and pleasant grasslands.

Sturt also noticed that, 'The grass consists of *Panicum* and several new sorts, one of which springs from the old stem. The plains were verdant indeed, the luxuriant pasturage surpassed in quality ... anything I had ever seen.'[36]

Mitchell, Sturt, and others spoke of the prodigious growth of oat grass, and how much their stock and horses enjoyed and prospered on the feed. Today we refer to that grass as kangaroo grass, and it is the mainstay of almost every 'unimproved' pasture in the country. In the past, its seed would have been a boon to the Indigenous inhabitants. Contemporary Aboriginal residents of Cooma report that their horses will gallop past all other grass to eat the seeds of *Panicum*. These grasses have enormous agricultural potential.

Another plant, Coopers clover (*Trigonella sauvissima*), had also been grown and harvested by Aboriginal people, but was also favoured by introduced stock. Its loss impacted severely on Aboriginal economies. Mitchell first saw it in a lakebed:

> [It was] at this time covered with the richest verdure, and the perfumed gale which … heightened the charm of a scene so novel to us. I soon discovered that this fragrance proceeded from the plant, resembling clover, which we found so excellent as a vegetable during the former journey.[37]

The loss of this dietary component was crippling to the Aboriginal economy because, as with the oat grass, yam, and nardoo, introduced stock zeroed in on these plants wherever they grew, and as a consequence the Indigenous people lost both their habitation sites and one of their principal forms of sustenance.

Settlers were assiduous in preventing Aboriginal people from returning to these locations, the richest areas of Australia. But one Wimmera settler noted that after three or four years of grazing, 'Many of our herbaceous plants began to disappear …the silk grass began to show itself in the edge of the bush track … the long deep rooted grasses … have died out.'[38]

Nardoo (*Marsilea drummondii*) was a crucial plant because of its ability to grow on the beds of shallow lakes in otherwise inhospitable regions. As the lake dried, explorers observed Aboriginals sweeping the seed into vast stockpiles and processing it into flour, the excess of which was stored in a variety of vermin-proof vessels. Many explorers, including Giles and Ashwin, survived difficult stretches of their journey only after plundering these reserves.

The grain harvest supported populations so large that many reading this for the first time will be amazed. This is our great desert, the dead heart of Australia — it goes against our mythology, which insists on its hostility to humans. Our poets laud the emptiness as a psychological marker for all Australians.

As recently as October 2013, an article in *Australian Geographic* about Australian deserts talked about the terror at the heart of Australia's geography, but managed not to mention Aboriginal use of the land at all, even though, as late as 1875, Lewis saw 350 people in these regions, and others had seen 500 or more.[39]

Good anecdotal evidence comes from the Australian author Mary Gilmore, who recorded the memories of her family, some of the first settlers in the New England region of New South Wales. Her uncles recalled a similar ceremony to the Katoora (described above by Duncan-Kemp) as well as dam building, irrigation, and harvests.

And on the creek where Bourke starved to death, the explorer McKinlay noted, 'The whole country looks as if it had been carefully ploughed, harrowed, and finally rolled.'[40] Descriptions like this are common in the first colonial records; the settlers hardly had to fell a tree to begin grazing stock, but almost none credited these conditions to Aboriginal management.

Kate Langloh Parker was a writer from northern New South Wales, and one of the first to record Aboriginal stories. She described these harvests in great detail as late as 1905. Australia was already federated, but the Yuwaaliyaay were still harvesting grain by traditional methods. The barley grass was cut and thrown into a brush-fenced compound, and then set alight and stoked continuously so grain fell from the stems into a collection pit prior to threshing.[41]

When early settlers found an Aboriginal tool that looked like a hoe, it was dismissed because they had convinced themselves that there was no agriculture in Australia. If you're not looking at these tools with an open mind, they are considered aberrations. Alter your perspective by a few degrees, and the view is different. Robert Etheridge was palaeontologist to the geological survey of New South Wales and the Australian Museum at Sydney. In 1894, he speculated on the use of these 'hoes', and concluded that the myth that Aboriginal people had no knowledge of husbandry was a mistake based on prejudice.[42]

These implements have received very little study since that time, falling as they do outside Australian assumptions of Aboriginal achievement. Some commentators decided they were items of penis worship: the researchers assumed Aboriginal people were too backward to have cultivated the land, so the stones had to have a phallic significance. Australian scholarship has a variety of routes toward demeaning the first possessors of the soil.

After reading the first edition of *Dark Emu* (2014), Jonathon Jones, a young Wiradjuri artist and archaeologist, asked to examine the stone-tool collection at the Australian Museum, and there he found dozens of these implements. They were so large and heavy that he deduced they had to be used in a pendulum fashion between the legs, like a pick or plough. The marks left by the bindings indicated they had been fitted with a handle at right angles to the tool. The stone had not been used to work wood or stone, but had been worked only in soil. It is these implements that are crucial to our understanding of Aboriginal agricultural history. They have received so little study, most have never been displayed, and almost all have no labels. The few that are labelled are tellingly referred to as Bogan River picks!

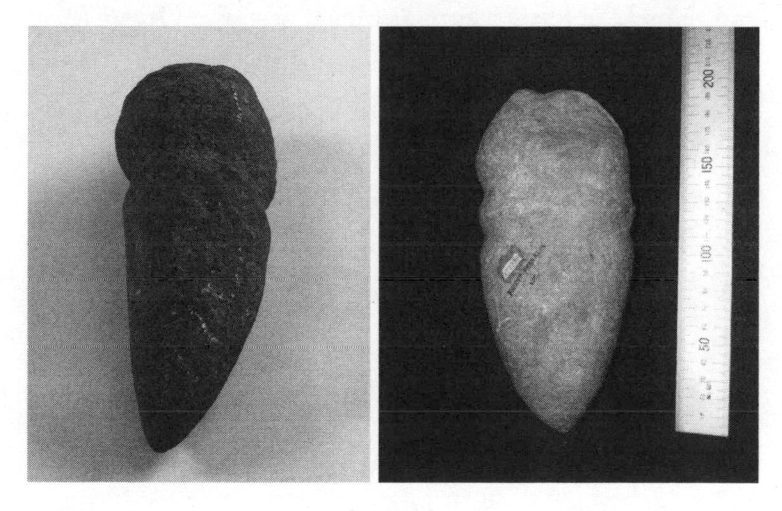

Stone picks

(Jonathon Jones)

Since the publication of *Dark Emu*, I have received photos from farmers around the country showing other strange implements that have been in family collections for over a century. A study of these tools will shine a fascinating light on the history of world agriculture.

Similarly, many northern Australian museums display long, knife-like implements, which usually bear legends such as 'of unknown use' when, in fact, they are juan knives — long, sharp blades of stone with fur-covered handles, which the explorer Gregory described the Aboriginal people using to cut down the grain.[43]

In 2010, I was shown a couple of stone implements found in the Colac region of Victoria. They were long, pear-shaped plates almost as large as the blade of a canoe paddle. I saw a similar object at Cape Otway (near Colac) in 1998, but the purpose of these objects is not known with any certainty; however, similar objects in Queensland and New South Wales have been found to be platforms for food preparation. Are they ceremonial objects, or kitchen tools? Only analysis of the surfaces can tell us definitively what they were used for, and so far, little analysis has been done.

Domestication of food plants

The scholar Rupert Gerritsen assembled a large body of material about the progression of people worldwide towards sedentism and agriculture. One of the tests for this progress is the domestication of plants:

> When plants become 'domesticated' as the result of a human induced selection regime, they undergo changes in form and structure to such an extent that they often become a new species. Genetic change takes place in this process and the subject plants become dependent on humans for the continuance of their life cycle.[44]

Some of the changes that occur in plants after domestication are reduced dormancy; a tendency to ripen simultaneously; and the development of a tough rachis around the seed, which inhibits germination unless associated with artificial watering. Harvesting and winnowing techniques also contribute to changes in seed characteristics. Just such qualities have been found in the initial, although belated, studies of Aboriginal grains. Gerritsen brought together the work of Zohary and others to show that Aboriginal people were performing the same cropping activities as those that led to the domestication of wild wheat and barley in Europe. These researchers claim that a tough rachis developed within just twenty to thirty years of this style of cropping, to the extent that it prevented germination without an artificial watering regime.[45]

Australian grains became dependent on the interventions of Aborigines, and the wide grasslands, monocultures of grain, were the result of deliberate manipulation by Aboriginal people.

Similarly the desert raisin, or bush tomato (*Solanum centrale*), used by Central Desert people for thousands of years, has become dependent on people for its propagation and spread. As a favoured plant, it is most commonly found near campsites, and is promoted by selective burning, further strengthening the reliance of the plant on human intervention.

Such is the plant's esteem in Central Desert culture that a good person is often described by its name. Custodians celebrate the plant in ceremonies, dance, and song, and body painting often features its image. Surplus harvests were ground to a paste and rolled into balls, which could be used more than a year later.

The dearth of research on these plants should not imply a lack of evidence, but rather a lack of interest and will to explore Aboriginal interaction with the plant communities of Australia. Some of these bush fruits and seeds are becoming popular in restaurants, but yams and grains have attracted very little scrutiny. The need to undertake research on the Aboriginal economy is not only necessary for the mature understanding of how Aboriginal people lived, but also for how and when human families spread across the earth.

We've been taught that Aboriginal peoples arrived in Australia after crossing land bridges from Indonesia. When I went to school, it was assumed that this happened 10,000 years before sea levels began to rise — after the Ice Age. After the introduction of carbon dating, that figure ballooned to 40,000 years and, after further research using more modern dating techniques, to 60,000 years. Aboriginal people have, of course, been saying we have always been here.

Western Victorian Aboriginal communities had been trying for thirty years to have an ancient midden on the Hopkins River analysed. It is so old, it has turned into rock. When that examination was finally undertaken, it was found that the midden was as old as 80,000 years, 10,000 years before the Out of Africa theory says humans began to leave Africa. As the Hopkins River midden is in the extreme south of the continent, it is probable that more ancient middens exist on the edge of the continental shelf, where people lived prior to the sea level rising.

It's an interesting proposition to investigate for our understanding of all world peoples and their movements across the globe. Professor Jim Bowler's geological work at Lake Mungo produced startling results, but his work on the Hopkins River has even greater significance. It is an exploration of the past that all Australians can enjoy, and an opportunity to further acknowledge Aboriginal Australia's exceptionalism.

In the last ten years, other researchers have begun intensive study of the plants domesticated by Aboriginal people, and we need to encourage and support this research.

As one of Australia's most senior archaeologists confided to me after struggling to gain official interest in her excavation of a sophisticated village site in the Murray River region, it is easier for Australian archaeologists to get research grants overseas than ones undertaking new areas of research in Australia.

Most archaeologists believe that the move towards sedentism is always associated with some form of agriculture, and is described as the period of intensification. Bill Gammage is of the opinion that both yam and grain harvests required some form of sedentism, but the 'power of Aboriginal spiritual sanctions' guaranteed that crops would not be interfered with by surrounding clans.[46] This precluded the necessity to stay by crops to protect them, and allowed greater opportunities to travel and to engage in prolonged cultural rituals.

Stories of ancestors teaching their people about selecting seed, sowing it, and building dams are common in the grain areas. Alfred Howitt, a Gippsland police officer and amateur anthropologist, records that the Dieri believed in mythological ancestors who distributed five major food plants. Peter Beveridge, from his property near Swan Hill, recorded the traditional stories of the Wati Wati, most of which have food production, soil preparation, and storage of the surplus harvest as a central theme.

Another grain cultivated and harvested by Aboriginal people was rice. These grains tended to be utilised above the Aboriginal grain boundary in the map based on Tindale's work. (See page 28.)

In July 2012, I heard Ian Chivers, adjunct professor

at Southern Cross University, discussing the potential of native rices on ABC Radio National. He stressed the importance of the genome of the Australian rice, because Asian rices were losing the characteristics that protected them against diseases, among other deleterious changes. In the conversation, Aboriginal people were not mentioned, so I rang him to discuss his research. He wasn't sure how involved Aboriginal people had been with the grain. They used them, but did they plant them?

Chivers was interested in the idea of Aboriginal grain management, but his focus had been on the economic potential for native grains in contemporary markets.

A book referred to me by Chivers' research partner, Frances Shapter, *Australian Grasses* by Fred Turner, doesn't mention Aboriginal people at all, but a more recent paper by Chivers argues that, 'In Australia we have stunning examples of very long term grain food production that had no degrading impact on the environment, that did not require expensive fertilizers or pesticides, and grew without the need for irrigation water.'[47]

Aboriginal people made changes to the genomes and habits of these plants simply through the continuous interference in the plants' growth cycle and selection of seed for harvests. This process, conducted over long periods of time, is what scientists call domestication.

In his contribution to *The Conversation* website, Chivers writes, 'These long-term cereal production systems were a feature of Aboriginal-Australia farming systems for thousands of years.'[48] Furthermore, he suggests we should be:

> looking at perennial grasses for our grain types, not annuals ... Can you imagine a permanent pasture that also produces a grain crop in those years when the rainfall amount and timing permits? It would also be the pasture that is able to survive the drought that will inevitably occur without the need to re-sow once the drought breaks ... This is a perennial grain-cropping system as it was used in the long-time past but which is still there for the discovery if we are wise enough to look.[49]

On 19 October 2012, Drs Penny Wurm and Sean Bellairs from Charles Darwin University took the discussion further on ABC Radio's *Rural Report*. They remarked that rices had been used by Aboriginal people for thousands of years and that, among other advantages, some species could be grown in brackish water.

These native rices were comparable to commercial rice and had a reddish colour, which they thought might appeal to culinary experts:

> Native Australian rice has been harvested and consumed by Indigenous people for thousands of years (and) may have the potential to underpin a wild rice enterprise as a 'bush tucker', 'novelty' or gourmet

product (either as grains or flour) for the tourism and niche gourmet markets.[50]

Like Chivers, Wurm and Bellairs believe these wild rices are a significant genetic resource for developing cultivars and for economic development in the Asia–Pacific region.

This is good news, and hopefully it means Aboriginal people will be able to take advantage of the intellectual property they have invested in this plant. Sun Rice Pty Ltd assisted in the research, and, as they stand to benefit from research results, I hope they'll respect the contribution that Aboriginal people made to the research over thousands of years.

Irrigation

Many explorers and pastoralists saw dams and irrigation trenches, but Walter Smith also saw them being built. Kimber records a conversation with Smith:

> The people would get in a line, using their digging scoops and larger coolamons. The clay and earth was scooped into the larger coolamons, which were passed along the line. Walter commented on the speed of the operation; with a line of people working the deepening of the favoured catchment area and the building of the bank could be done at the same time. When it

> was satisfactorily excavated, the people would trample the clay base. If ant nest material was nearby this was carried and trampled in to give a very firm base.[51]

A dam wall discovered on the Bulloo River floodplain in the Channel Country of south-west Queensland was 100 metres long, two metres high, and six metres at the base; it required 180 cubic metres of material to construct. The clay was mixed with gibber pebbles to create an earthen embankment across the catchment of several streams, and was capable of holding 185,000 gallons (700,000 litres).[52]

Near the Gregory escarpment, in the Gulf of Carpentaria, there is an outstanding example of agricultural engineering: 'It is a series of interconnected dam walls creating many permanent deep pools that makes the whole complex extraordinary.'[53] This technique was said to enhance the yield obtained (a similar method is reported by Kimber, 1984). We are not talking about ambiguous one-off examples. These structures were seen across the continent.[54] Norman Tindale saw a dam on the Nicholson River in 1977 that was designed so water would spill over and irrigate grainfields. Kimber wrote about similar constructions: Godfrey's Tank in the Great Western Desert was named after one of David Carnegie's men. The site is surrounded by art and carvings, and was estimated to have held over 40,000 gallons (151,000 litres).[55]

Wiradjuri people in New South Wales also built large dams, and then carried fish and yabbies in coolamons over large distances to stock the new waterholes.[56]

In 1875, Giles found a dam near Ooldea, South Australia (on the eastern edge of the Nullarbor Plain); it had a bank 1.5 metres high, and was 1.5 metres at the base. An overflow channel to allow floodwater to spill away without damaging the wall had been built on one side. Giles thought of the works as crude, but, 'for a full week it watered seven men, 22 camels and filled up enough water containers to last them on a dry stage of 500 kilometres'.[57]

An extraordinary construction was found by early explorer S.G. Hubbe when he was looking for a stock route into Western Australia.[58] A clay and granite wall of 1.8 metres was built at the base of a granite outcrop. As the soil at the base of the dam was friable, it had been faced with granite slabs. Steps had been cut into the dam wall, and the catchment was so good that even the slightest shower would result in a substantial collection of clean water.

Sturt also found a large well north of Lake Torrens, South Australia, which was:

> 22 feet deep and 8 feet broad at the top. There was a landing place … and a recess had been made to hold the water … Paths led from this spot to almost every point of the compass, and in walking along one came

> to a village consisting of nineteen huts ... Troughs and stones for grinding seed were lying about ... The fact of there being so large a well at this point (a work that must have required the united labour of a powerful tribe to complete) assured us that this distant part of the interior ... was not without inhabitants.[59]

Archaeologist John Morieson and students from Swinburne Institute recently investigated numerous wells at Kooyoora National Park, near Bendigo, which, when cleared of accumulated debris, would fill to overflowing after receiving only a few millimetres of rain. A few were capable of holding thousands of gallons.[60]

Wells at Kooyoora National Park, near Bendigo, Victoria

(Lyn Harwood)

Morieson and others investigating the construction of such wells speculate that they were deepened with the aid of repeated exfoliations of stone after fires were lit in the well, and that cold water was then applied to the hot rocks, causing large flakes of stone to lift away from the sides and bottom. Others suggest that even the natural decomposition of the stone and continuous scraping over thousands of years with stone trowels would contribute to the deepening.

There were numerous methods employed to take advantage of the available rainfall. In the north-west goldfields of Western Australia, people sowed the seeds of kurumi (*Tetracornia arborea*) in cracks in the clay pans to ensure propagation in the wet season. They created a stone arrangement on the bank to record the story of the collection, preparation, and consumption of the harvested kurumi.[61] Morieson records similar baffling stone arrangements in Victoria, which may have been created for the same purpose.

Settlers and explorers from other areas have reported large well systems, miles of stream diversion, and systematic flooding to prepare the ground for sowing seed. However, as soon as such wells were discovered, they were commandeered by sheep and their shepherds, because they were situated close to the croplands to which the sheep gravitated with unerring accuracy. (Settlers were not droving stock into new areas, so much as simply following on behind until the sheep found the next crop or garden.)

It seems plausible that scientists could examine the

origins of some of these dams and water-storage systems to prove the involvement of Aboriginal people. A simple probe into the banks of dams situated on old grazing properties might register different construction techniques below the surface. Barber and Jackson have already applied similar techniques to the Roper River in the Northern Territory, where Aboriginal dams were frequently observed in the early post-contact period.

People near Elsey Station in the Northern Territory created dams to maintain fish ponds. This brought them into conflict with the owner of the station, who wanted the water for the exclusive use of his cattle. The police and courts backed the station owner, Holt, and the people were punished. But their greatest punishment, of course, was the loss of their resources and livelihood.

It is time we examined Aboriginal water conservation with a more generous attitude. Further scientific surveys in other districts have the potential to uncover complex systems in other geographies and weather zones; in a drying continent, this information will be important, and could bring about change in Australian perceptions of Aboriginal engineering, conservation, and labour division.[62]

Game or farm

Game hunting employed a much more reliable procedure than opportunistic kills in the field. Some of the earliest explorers noticed various interventions carried out by

Aboriginal people when hunting. They saw massive poles that had been erected on opposing banks of waterways, and speculated on the use of these large, robust constructions. Later explorers were present when Aboriginal people strung fine mist nets across streams from such poles to catch ducks and other fowl. The nets were only in operation for short periods so as not to interrupt the ducks from following their natural path along the waterways. Mitchell wrote that:

> The meshes were about two inches wide, and the net hung down to within five feet of the surface of the stream ... Among the few specimens of art manufactured by the primitive inhabitants of these wilds, none come so near our own as the net, which, even in quality, as well as the mode of the knotting, can scarcely be distinguished from those made in Europe.[63]

The pastoralist James Dawson, and Robinson, mention game drives or 'grand battues' where people were engaged in driving game across a 32-kilometre front to a dispatch point. Dawson reports on the co-operation of several tribes involving over 2,000 participants. One settler in Queensland found a 'Kangaroo net fifty feet long and five and a half in width, made of as good twine as any European net.'[64]

Colonists witnessed these nets used in combination with kilometres of brush fences in large-scale trapping or battue operations. Remnants of the walls, outlining wings of these battues, can still seen in some parts of the country. Near Euroa in Central Victoria, a massive system of stone walls connects natural rock outcrops. Lichen growth on the rock is claimed by some botanists to indicate that they were constructed well before the period of contact with early European settlers. This particular drive brought kangaroos from a huge flat area to the foot of the range, and then chuted them into a series of holding pens where narrow apertures could direct animals designated for slaughter one way, and those to be released in another.

The stone works and nearby housing associated with these drives represents an incredible labour investment and a move towards sedentism comparable to that represented by fish traps at Lake Condah and Brewarrina. Sites such as these are begging for further investigation.

One researcher claims that the Garden Range Aboriginal rock-art site in the Strathbogie Ranges of Central Victoria, close to the battues at Euroa, depicts the activity of herding and farming kangaroos. Nonetheless, most of the tool workshops associated with these constructions, as well as the constructions themselves, still do not appear on the archaeological register of Aboriginal Affairs Victoria.

Michael Archer of the Australian Museum puts a case for the use of kangaroos as a food and farm species.

He and other scientists speculate that the culling of kangaroos by Aboriginal Australians had little impact on their populations, because adult males were targeted. In demonstrating the sustainability of the industry, Archer quotes studies on Mulyungarie Station in South Australia where the harvesting of 10,000 males a year over eight years saw an increase of animals from twenty to fifty per square kilometre.[65]

Kangaroo flesh has a low fat content, and is free from impurities, as the animals do not require chemical drenching. They can tolerate a harsh environment and, moreover, their feet do not break up the surface of the soil, or compact it — both of which lead to erosion.

The Aboriginal battue system of kangaroo- and emu-harvesting suggests a way of drafting animals without the need for shooters, and this method may appease those city voters with an emotional attachment to our national emblems. Interestingly, the battues were associated with constructions that appear to be designed for predicting the solstice, a knowledge that Aboriginal people involved in the cultivation of plants would have needed.

Strangely, though, when Ross Garnaut, who prepared the climate-change policy for the Rudd Government in 2008, championed kangaroo farming as a way of conserving the land and cutting greenhouse gasses, because cattle are greater polluters than motor cars, the press could hardly contain their contempt.

Still not convinced?

The sorts of activities described above were not isolated examples. Large numbers of people engaged in various agricultural activities were observed throughout Australia. Near Cooper's Creek, one settler saw women collecting seeds and roots on the flats, 'as thick as grazing sheep'.[66] Peter Latz, a central Australian botanist who grew up on Hermannsburg Mission, described the technique of women harvesting the tubers of onion grass (*Cyperus bubosus*): 'Women sometimes dig a trench at the edge of the patch then work in a line, turning over the ground as they go.'[67] Many early explorers witnessed this activity, and recognised the efficacy of its cultivation process.

King, on the doomed Burke and Wills expedition, found a store of grain in an Aboriginal house, which he estimated at four tons. John Davis, a member of one of the search parties for Burke and Wills, reported on the vast quantities of nardoo seed waiting to be harvested on the dry floor of Lake Coogiecoogina in the Strzelecki Desert, reminding us that 'desert' is a term Europeans use to describe areas where they can't grow wheat and sheep.

Howitt, on another search party for Burke and Wills, also found large stores of nardoo. Early settlers on the Mulligan River in Queensland remarked on vast quantities of nardoo being harvested, and the explorer/drover Ashwin, not prone to giving credit to Aboriginal people, found two granaries, 'one with about a ton of rice seed stored there in 17 large dishes'. His comment on this find of 'delicious grain' was that it was 'a pity we did not take more'.[68]

Houses, water races, harvest fields, and irrigation may have been observed, but within weeks, sometimes hours, of observation, fire destroyed the houses, sheep and cattle destroyed the fields, and the dams were usurped for European use.

The explorers' journals are full of their surprise at finding evidence of Aboriginal utility of the land. Apart from the tubers, grains, and fish, as reported by Walter Smith, Bill Harney[69] and others, they also described the herding, corralling, and harvest of young water birds of various species.[70]

Innumerable commentators came across the preservation of everything from fish, game, plums, caterpillars, moths, quandong, figs, seeds, and nuts, among a wide variety of other foods. Preserved caterpillars were made into a kind of flour; figs and quandong were pulped and mixed to form a product that can only be likened to quince paste.

Sir Joseph Banks didn't like the fruit of the Australian banana (*Musa acuminata* ssp *banksii*), but it was the trunk the people ate after cooking it on hot stones. It tasted like green bananas, and the plant immediately re-sprouted from the cut stem.[71]

Kirby and Beveridge found vast acreages of rushes that the Wati Wati were harvesting and nurturing: 'The reeds looked like large fields of ripe wheat; and nearer, where they had burnt them, it had the appearance of a splendid crop just before it comes into ear.'[72] This was a managed system, and the management had produced a scene familiar to European eyes.

Kirby described the meal from this compung (cumbungi) rush as very similar to flour or potato meal.[73] Mitchell said that the cakes made from the cumbungi flour 'were lighter and sweeter than those made from common flour'.[74] Huge mounds were raised in these reed marshes near Swan Hill so that villages could be located at strategic locations within the swamp to manage the harvest of this valuable plant. During their first days in the district, both Kirby and Beveridge were intrigued by these massive mounds and the fact that they were emitting steam. Upon examination, the mounds proved to be gigantic ovens for the cooking of the compung rush.[75]

Mitchell noted 'the lofty ash hills of the natives, used chiefly for roasting the "balyan" (or bulrush)',[76] and recorded how astonished he was by the volume of starch produced.

The base of this plant, if used fresh, is like the freshest, crunchiest salad vegetable you've ever tasted. Explorers Eyre, Kreft, and George Moore all refer to the importance of bulrush starch in different parts of the continent.

In his later years, Beveridge recalled many of the traditional stories of the Wati Wati, and, despite his appalling handwriting and the fact that he wrote over other printed material, the striking thing about these stories is that so many have planting, husbandry, harvesting, and storage as central themes. The ancestors had left instructions for the care of plants and the sharing of the produce. Some stories are so specific in their instruction that recipes are given.[77]

The instructions for mallee-fowl eggs in the story 'Coorongendoo Muckie' of Balaarook, near Swan Hill, go as follows:

> When Weitchmumble had secured all the eggs the Ngalloo Watow set to work to make fire by rubbing a narrow lathlike *(sic)* piece of saltbush across a sun crack in a pine log. A few minutes of rapid friction were sufficient to perform the operation, therefore the camp fire was soon made and half of the lowan eggs set on end in the sand before it were in a short time simmering away, being stirred the while with a thin twig, through an opening at the top, made for the purpose, and when cooked they presented the appearance and consistency of a yellow paste, and, as to the taste thereof, the adjective talke (good) conveys but a very remote idea.[78]

Gerritsen quotes a similarly vivid recollection by the German missionary Johannes Reuther at Lake Killapaninna, South Australia, about the ancestral being Markanjankula:

> At first he came to Aruwolkanta (wara ngankana) where he found a beautiful level plain. Here he cleared

> away all the weeds and stones, loosened up the soil, sowed some ngardu [nardoo], and then covered up with earth what had been sown, so that should a flood come along, the ngardu should come up.[79]

Markanjankula is referred to in other parts of the legend as broadcasting seed, and digging out and creating seed-grain pits.

It seems that even the name of Lake Killapaninna has within it the word for the harvest grass variously spelt as pannana or parrara. The evidence, while now difficult to find on the surface of the land, is still embedded in the language.

It may be that not all Aboriginal peoples were involved in these practices, but if the testament of explorers and first witnesses is to be believed, most Aboriginal Australians were, at the very least, in the early stages of an agricultural society, and, it could be argued, ahead of many other parts of the world.

In an article in *Antiquity*, Denham et al. say, 'If the dispersal of the greater yam occurred before the separation of New Guinea and Australia … then horticultural experimentation occurred in northern Australia at least 10,000 years ago.'[80]

Norman Tindale, the early-twentieth-century Australian anthropologist, archaeologist, entomologist, and ethnologist, after examining irrigation and horticulture and the stone technology that supported them, estimated the milling techniques to be around 18,000 years old, an age which, if it is true, re-writes the history of world agriculture.[81]

When I came across Tindale's assessment of the Australian Aboriginal tool kit, it seemed at odds with much of what had been written on the subject, and I could find little to support it. However, opinion has changed in the last decade, and Tindale's conclusions no longer seem so outrageous:

> The ground edge axe ... has been invented, presumably in Australia ... As the antiquity of edge grinding and axe shaping techniques, such as pecking, is pushed further and further back it is apparent that their diffusion took place within the greater continent of Sahul Land rather than from across the seas to the present Australian continent.[82]

Stone-tool studies in Australia are often used to prove one theory or another, but, much as we should treat Tindale's theories with caution, we must also be careful about accepting the views of some current archaeologists who maintain that Aboriginal stone tools underwent an accelerated technological advancement, or intensification, 4,000 to 5,000 years ago. Many sophisticated Australian tools have been tested and found to be of much greater age than this, and would seem to resist the idea that intensification only began in Australia 4,000 years ago.

Importantly, a recent survey in Western Australia by Sue O'Connor from the Australian National University has uncovered an edge-ground axe almost 50,000 years old, easily the oldest of its kind in the world.[83] When Norman Tindale suggested that this might be the case, he was ridiculed; now it seems we need to look at the Aboriginal technology without the assumption of its poverty.

Examination of Aboriginal tool kits show how closely they are fitted to the economy. That is not so surprising, but it is revelatory that the technology has been found to be much more closely associated with crop utilisation than previously thought. Further studies are essential to our understanding of how Aboriginal people interacted with plants, and the extent to which they were in control of that interaction. Was their use of the plant opportunistic, or more closely aligned to agricultural practice? Later examples in this book reveal the methods and technology that were developed to gain greater and more reliable yields.

We are at the beginning — not the end — of understanding pre-colonial history, and the most recent archaeological research suggests very old ages for Aboriginal occupation. Wade Davis, when analysing the evidence for early Aboriginal occupation, used the figure of 60,000 years ago for when modern people began to leave Africa. If that is so, then the generally accepted figure for Aboriginal occupation of Australia of 60,000–65,000 years ago makes us one of the first, if not *the* first, to leave the African continent.[84] As suggested earlier, even more recent discoveries have put that age beyond 80,000 years.

Every month sees new archaeological information put before the public. As we have seen, the Warrnambool midden has been dated at around 60,000–80,000 years ago, a site in Kakadu at 65,000 years and a cave in the very driest part of South Australia has shown Aboriginal occupation of around 50,000 years — much earlier than it was previously thought Aboriginal people had occupied that region.[85]

The need for further examination of ancient Aboriginal influence on Australian plant communities and landscapes has been stressed by A.P. Kershaw, the environmental scientist whose survey of site 820 was designed to study the human presence in Queensland. He claims, 'The weight of available evidence points towards Aboriginal burning as the most likely cause of vegetation changes … and this implies that people have been present on the Australian continent for at least 140,000 years.'[86]

That this claim is contentious is obvious, but earlier pollen-core studies by palynologist Gurdip Singh at Lake George, near Canberra, show similar activity resulting in sudden changes in land use. Singh proposed that this dramatic change in vegetation was a result of Aboriginal fire-stick farming. Eric Rolls discusses these findings and the implied potential for an early arrival of humans on the Australian continent in an unpublished manuscript.

The term 'fire-stick farming' was first proposed by archaeologist Rhys Jones in 1969, but more recent studies have increased the understanding of Aboriginal land use, and support observations by the explorers. This indicates that the trajectory towards agricultural activities may have begun much earlier than we currently believe. It's interesting that Jones chose to use the word 'farming' in the term he introduced more than forty years ago.

Studies of fish traps on the lower Murray River and the Coorong Wetlands by the maritime archaeologist Peter Ross led Jones to conclude:

> Adaptations that had been attributed to intensification during the late Holocene have been found to occur as early as the Pleistocene in some regions … Subsequently, archaeology in Australia is currently reconsidering regional findings that appear to be at odds with the continental narrative.[87]

The 'continental narrative' is one in which a change in the economy of peoples occurs because of external forces, such as a change in the environment, not, as is postulated by anthropologist Harry Lourandos, as a result of social change. It is this accelerated change in technology and behaviour that archaeologists refer to as 'intensification'.

Scientists world-wide believe that human engineering and agricultural experiments began around 4,000 years ago, and that tool developments were naturally associated with this change. The evidence in Australia, however, suggests that those changes may have begun here much earlier.

Harry Lourandos says, 'The old distinction between "resourceful" agriculturalist and "quiescent" hunter can no longer really be seen to apply.' If we examine the past, it cannot support the idea of 'passive adaptation to changing natural environments, but [rather] active participation in complex interplays — among them, social, environmental and demographic'.[88] Aboriginal people were not reacting to the state of nature, but directly affecting its production. 'Aboriginal culture has been changing and expanding over a long period of time. The more recent changes of the last 206 years are simply a continuation of a tradition which goes back thousands of years.'[89]

Australian farms of the future

Farmers have always been crucial to land conservation, and have always had a more practical approach to soil conservation than most of their critics, but it is the reliance on European plant and animal domesticates that has caused them the most conflict with this continent.

What would happen if we turned away from total reliance on sheep and cattle, and diversified into emu and kangaroo? Researcher Gordon Grigg argues that:

> [G]raziers already run sheep on top of kangaroos (and other herbivores) and the grazing pressure is too high.

> If they make money from kangaroos, and if kangaroos become accepted as an economically positive part of their mixed grazing system, they will at least have an option of maintaining economic viability with lower sheep numbers.[90]

One of the problems that prevents acceptance of kangaroos as a resource is ownership. As Archer says, 'Conservationists are concerned about the principle of giving up ownership of wildlife to private citizens. Graziers are concerned about their lack of ownership of a resource on which I am suggesting they should become reliant.'[91] The utilisation of those animals that are already adapted to our climate and geography, and that damage it less as a result, should become the subject of serious ecological and economic debate in Australia.

Similarly, what would happen if we tried some of the Aboriginal grains instead of the thirsty and disease-prone grains of Asia and Europe? After studying Aboriginal yields from yam daisies, it is easy to imagine a potato farmer turning over part of his farm to yam, thus avoiding the need to use fertiliser and herbicides.

The murnong (*Microseris lanceolata*) is sweet and crisp, and metabolises sugars in a way that is much healthier for our bodies than are many current commercial crops. The juice produced during cooking — minne in Wathaurong language — is dark and sweet, and you can imagine how it could complement a curry. Most of the

Australian grains are gluten free, and don't require heavy chemical supplements for their successful cultivation. Farmers dedicated to their sheep, cattle, and wheat need do nothing, but more adventurous agriculturalists might welcome the challenge.

Mitchell grass, Brewarrina, northern New South Wales, 2010

(Lyn Harwood)

Rolls spruiked the virtues of *Astraleba lappacea* (Curly Mitchell grass) with six-inch long ears and filled with clean, firm grain. Archer urges us to consider also *Panicum decompositum* (Mitchell's native millet) and *Themeda avanacea* (native oatgrass), and Chivers' research encourages examination of native rices. Researchers think there are over 140 Australian grasses that were harvested by Aboriginal people. The Gurandgi Munjie

Food Group's involvement in grain and murnong field trials have facilitated the discovery of other grasses such as *Themeda avenacea*, and *Sorghum leiocladum* (native sorghum), which show enormous potential as plants useful to the Australian diet and economy.

There's no contemporary market for these grains, but I bet a stall in any city market could sell flours from these grains at premium prices to whole-foods enthusiasts. Markets are created by entrepreneurs. Set aside a few paddocks and have some fun, and I'll eat my boot if it doesn't yield a profit.

Archer lists a whole series of bush fruits and plants that are ripe for enterprise. We seem to be stuck on wattle seed and lemon myrtle, but supply a few good cooks and whole-foods shops with other top produce, and watch as Australia transforms its diet and develops a genuine Australian cuisine.

Archer compares the land degradation of farming, which occupies 70 per cent of the land, to mining, which takes up only 0.02 per cent He suggests that mining produces high yields for relatively low costs, whereas farming is a high-cost, low-yield enterprise. Any efforts we make to conserve our land would best be spent encouraging farmers towards more soil- and bank-friendly activities. Australia must continue to produce food, and farmers are the heart and soul of this enterprise, but there have to be better ways to farm the light Australian soils.

The great advantage of Aboriginal crops is that

they have been developed through seed selection, direct planting, and weeding for the harsh conditions of Australia. Many of the grains grow on sand, and require a minimum of irrigation. The good news is that the Rural Industries Research and Development Corporation has been studying some of these grains with a view to incorporating them in the modern agriculture of Australia.

Latz says that 'the nutritional value of the seeds from the desert species is equal to or better than that of the cultivated grains'.[92] These indigenous plants promise a huge economic bounty for the country, and our future prosperity demands they be given serious consideration.

2
Aquaculture

Aquaculture was well established in Australia long before the first colonists arrived, and the examples in this chapter demonstrate and expand on the nature of Aboriginal intervention in food production across the country. James Kirby, for example, saw the automatic fishing machine near Swan Hill; Colac's first European settler, Hugh Murray, was delighted by the flavour of the whitebait in Colijan nets he stole at Lake Colac; and missionary Joseph Orton witnessed the Colac people fishing for whitebait, noting how they scorned the beads and mirrors left to replace fish taken from the nets by settlers. The Lake Condah aquaculture complex was obvious to all who saw it, and Thomas Mitchell witnessed the massive fish traps on the Darling River at Brewarrina, which some claim are the oldest man-made structures on earth.

Andrew Beveridge witnessed Aboriginal people on the Murray, at the Swan Hill settlement of Tyntynder in north-west Victoria, using haul nets with reed floats to keep the net at the top of the water and clay weights to keep the other edge on the bottom. The weights had been baked, like bricks, in the fire. Beveridge was astonished at the amount of fish caught. In other parts of the stream, permanent fences of stone, brush, and stakes were fixed in a zigzag fashion across the stream, with apertures to allow the passage of fish. Nets were fixed in these spaces whenever fish were required.[1]

Beveridge also remarked on a series of dykes placed across the Murray River floodplain to prevent it from receding too quickly during summer and thus ensuring retention of fish stocks. The dykes, built from vast quantities of clay, were over a metre high, and extended along the river as far as the reedy plains extended. The warm, shallow water in these weirs created perfect conditions for breeding fish.[2]

In the south-west of Western Australia, the Barragup fishing weir on the Serpentine River was described by Jesse Hammond circa 1860:

> A wicker fence was built across the stream, completely closing it from bank to bank, except in the centre, where a small opening was left. Through this opening a race was constructed by driving two rows of parallel stakes in the riverbed. The bottom of the race was

> filled with bushes, until there was only about eight inches of clear water above the bushes for the fish to swim through. On either side of this race was built a platform, about two feet six inches below the top of the water. On these platforms the natives stood to catch the fish as they swam through the race.[3]

On the opposite side of the country, near Purumbete in Victoria, there's a near identical structure shown to me by Trakka Clarke. Presumably, this design has been passed across the continent along the cultural and economic songlines.

Specialist nets were also used across the country for particular fish and crayfish, and required skill and patience to construct. Some of the nets took experienced net-makers three years to complete, and were up to 270 metres long. Sturt saw a 90-metre net across the Darling River, 'of the very finest craftsmanship'.[4] Hume also observed intricate and extensive net-making on the Darling River.

Uncle Max Harrison describes a vast fish trap in a bay near Bermagui. Now silted over, the trap comprised massive boulders, which were moved into position using long poles lashed to the stones — the natural buoyancy provided by the full tide helped move the boulders to create walls. This incredible system so far has received almost no attention. The local Yuin people are appealing to government to reclaim this system as an employment opportunity for their young people, not just in the re-establishment of the trap, but also for the fishery and tourism potential.

Robinson remarked on the incredibly successful operation of a fish trap at Pambula.

The Yuin also had a whale fishery at Boydtown, just south of Eden, where the people adopted the European tools and boats into a tradition of hunting whales that had been operating for hundreds, possibly thousands, of years.[5] Ritualised interaction with killer whales encouraged the mammals to herd larger whales into the harbour, where they would be driven into shallow water and harvested by the Yuin, who would then share the feast, not just with neighbouring clans, but with the killer whales themselves, who would receive the tongue.

The Yuin set up this interaction with the killer whales with a ceremony where a man would light two fires on the beach and pretend to limp between them as if he were old and frail. The Yuin believed that this encouraged the whales to take pity on the man and bring the bigger whales to the bay for his use. Europeans and Yuin combined to continue this operation for many years after first contact. It ended when a European man shot the lead killer whale. Unfortunately, the association between man and whale was broken in that instant.

Foster Fyans, a police magistrate, saw Aboriginals fishing at Geelong in partnership with dolphins who drove the fish in to the shore. Similar relationships were reported at Moreton Bay and many other Australian beaches.[6]

Various fishing methods were reported from across the country. John McDouall Stuart came across people fishing at brush weirs in the harshest parts of the country.[7]

The Condah system of massive eel concourses in south-west Victoria must have taken centuries to refine. The stone was readily available, but there was so much of it that great viaducts had to be created among them and channels chiselled through rock and earth — a vast and daunting operation, even if it were replicated today using modern machinery.

Brewarrina

The Brewarrina fishing system in the north-west of New South Wales is an example of a large-scale fishing operation, but it also reveals the economic and social organisation needed to sustain the fishery.

Fishing in the Brewarrina fish traps (See textual reference on p. 75.)

(Powerhouse Museum)

The fish-trap system is so old that the local Aboriginal people, the Ngemba, attribute its construction to the creator spirit Baiame. It is hard to get much information on this incredible construction but at an Aboriginal Languages meeting in Sydney in 2012 I met Brad Steadman, an elder from Brewarrina, who, upon hearing of my interest in the traps, told me one of the traditional stories:

> Bunggula, the Sooty Grunter (bream), grunts when taken out of the water. The spines on its back are the spears flung by the old man, Baiame, who hunted him in the waterhole. The fish escaped, and as he flashed his tail he made a channel which filled with water to make the river. But the country dried out, the kangaroos went away, the plants died, and there was a big drought. The old man came back with his dogs and his sons, and said the drought was because the people didn't know the law or the names of the rivers. He told them the songs to sing and the dances to dance so the rain would fall again and things would be as they are today.

The river is supposed to be only 5,000 years old, but the old people have stories from long ago that say those rocks (volcanic) were always there. The Bunggula turned up those rocks as he escaped from the old man's spears. The old man told the people how to arrange those rocks into the patterns they are in today.

Witnesses who saw the system in operation in the early 1800s were astounded by the efficiency of the traps, the efforts employed to maintain breeding stock, and the enormous harvest. Large numbers of people depended on fishing traps along most inland rivers, and the Brewarrina trap was only one of hundreds of such systems. The engineering of the water races and pounds was ingenious, and observers were amazed at how the structure withstood the regular floods. A stone locking system was engineered to fix the trap to the bed of the stream. The arch and keystones were two elements contributing to their strength.[8]

When I visited town, the curator at the Brewarrina Aboriginal Museum wondered aloud about the logistics of catering for the 5,000 and more people who attended the harvest each year. (The caches of milled flour that were pilfered by early explorers had probably been prepared for such events.)

It's hard to get much statistical basis for the claim that the Brewarrina traps are the oldest human construction on earth. There are only a few texts available on the traps, one being a three-page essay based on Barry Wright's paper presented at the New South Wales Aboriginal Health Conference of 1983. An archaeological team calculated the age of 40,000 years, but considered that to be a minimum, given the nature of their analysis.[9] The traps have been listed under state and national heritage lists on the basis of their antiquity.

Jeanette Hope and Gary Vines in their 1994 survey speculated that the traps were most likely built during periods of low water level, and this might be 15,000 to 19,000 years ago, or as recently as 3,000.[10] At any rate, they are very old, and rank among some of mankind's earliest constructions.

The Brewarrina Aboriginal Museum gave me an overview of the available research, and led me on an inspection of the traps and the museum exhibits. One of the most stunning photographs of the traps in use shows two young Ngemba men carrying fish from the traps. One of the fish is almost a metre long, and has a distinctive swallow tail. Since that time, no fish of this variety has been recorded in the river. No doubt, changes to the course of the Darling River to clear the way for paddle steamers, and the damming at many points, have interrupted the life cycles of more than one species of fish. Steadman told me that the fish in the photograph is birrngi, a type of bony bream that has now been lost (see photograph on p. 72).

The traps were also designed to allow the passage of breeding stock to pass through so that upstream fisheries could gain a share. Particular ponds in the system were managed and used by particular families — but those families had responsibilities for the secure provision of fish to the families and systems upstream and downstream from their location. It was an integrated and sustainable system.

Historian Peter Dargin wrote a book on the fish

traps for the Brewarrina Historical Society in 1976. It's a treasure, because it combines information about Ngemba belief systems with the most comprehensive technical data available at the time. It was Dargin who, paraphrasing R.H. Mathews, described the system of locking the boulders in place so that floods could not wash them away. More detailed information about that technique would allow us greater insight into the engineering skills of the Ngemba.

Dargin included wonderful drawings and photographs from the early-contact period; these are crucial to our understanding of the hydrology, given that more recent photographs show a system compromised by channels for paddle steamers, levelled areas for regattas, fords, and roads.

This wonderful little book is the sole champion for this ancient site. It is a mere seventy saddle-stitched pages. Gaffer tape binds it to hide the staples. The cover is all black, with reversed-out white type. It is impossible to produce a book more cheaply than this. Thank you to Peter Dargin and the Brewarrina Historical Society; without them, the antiquity of the fish traps might remain undescribed.

Rupert Gerritsen's important work was similarly bound and, for want of Australian interest, had to be published in London. Both his work and Dargin's are indicative of Australia's nonchalance about important considerations of Aboriginal culture.

There is still a lot to understand about Aboriginal

technology, and it would be highly significant if research upheld the claim that the Brewarrina traps are the first human construction. I hope it doesn't take us another 220 years to find out.

Early photograph of the Brewarrina fish traps

(Henry King)

Lake Condah and the Western District

The escaped convict William Buckley visited the Lake Condah traps before 1836, and extolled the quantities of fish they captured. He saw several other fish-harvesting systems on smaller streams throughout the lands west of Port Phillip Bay.[11]

Batman, as he began his venture into Victoria, saw fish traps on all the rivers he came across. He admired the ingenuity of the systems, and recorded their association with permanent housing.[12]

Morieson, in his manuscript describing Aboriginal stone arrangements, quotes Dawson's remembrance of such fisheries:

> Lake Boloke [Bolac] is the most celebrated place in the Western District for the fine quality and abundance of its eels; and, when the autumn rains induce these fish to leave the lake and to go down the river to the sea, the Aborigines gather there from great distances. Each tribe has allotted to it a portion of the stream and the usual stone barrier is built by each family with the eel basket in the opening.[13]

Morieson documents many of the fish traps in Western Victoria and the rather desultory examination of them by the Victorian Archaeological Survey. Although many have been destroyed by agriculture and rock collection for fencing, commercial purposes, and home gardens, a leisurely walk in the Lake Terangpom area reveals large embankments where nets and weirs were used to impound fish for the nutrition of a population that was largely sedentary. On the hill above this system is an ancient apple tree, which must have been planted by the European settler who, by whatever means, assumed ownership of the land. Those apple-eaters have gone now, their settlement a mere whim of the moment.

Square traps used as holding bays for fish stocks are readily seen at Lake Gnarput, Lake Corangamite, Lake Purrumbeet, and Lake Colac, but little work has been done to examine these systems. There has, however, been much study of the marsupial trackways at nearby Lake Milangil. These are extraordinary relics, easily the oldest and largest trackways in the world. Giant wombats and today's smaller cousin seemed to have coexisted in this area. This is fascinating by any measure, but within a kilometre there could be archaeological treasures marking the trackway of the human species.

Fortunately, humans will always have in their midst souls who cannot leave sleeping dogs lie. There are those whose minds unpick the seams of Australia's Bayeux tapestry to reveal what the victors tried to hide. Morieson's curiosity has urged him to tramp the Western District of Victoria for decades. He has charted the numbers of stone arrangements studied by the Victorian Archaeological Survey, some of which were listed in the National Estate but have since been destroyed by farming, malice, or ignorance. Investigation of the stone constructions of the Aboriginal fisheries and associated housing is imperative before all traces are obliterated.

Some Lake Condah fishery sites were seriously damaged after John Howard, the Australian prime minister at the time, panicked farmers into believing they'd all be ruined by Native Title claims. Morieson's catalogues and charts are valuable documents in view of the wanton destruction of so many sites.

Even at the time of writing, a massive roller was at work crushing volcanic stone in Western District pastures. On one level, it is simple pasture improvement; on another, it is heritage destruction. The operator of the roller is 'just doing what he is told', but he wouldn't be allowed to do it at Stonehenge or Easter Island.

Destruction of these systems was witnessed by the very earliest Europeans. Aboriginal Protector William Thomas saw many aquaculture systems, but reported that most were destroyed by Europeans in the first days after their arrival. One such system belonged to a particularly large village near Port Fairy, which had more than thirty houses capable of accommodating around 200 to 250 people. The whole village was burnt, and the sluice gates of its fishery destroyed.[14]

Another system, seen by settlers at Toolondo, near the Grampians in Victoria, connected two swamps with a 3.6 metre wide channel, which was over 1.2 kilometres long. Fishing platforms surrounded the system, with each section owned by a particular family.

Sue Wesson, in *An Historical Atlas of The Aborigines of Eastern Victoria and Far South Eastern New South Wales*, quotes Carmody's observation of fishing on the Murray:

> [Fish traps] were especially used at times of inter-tribal conferences, the Upper Murray being one of the centres where these large gatherings took place periodically. To

> ensure a plentiful supply of fish for the participating tribes, the trap would be closed with a special key stone twelve months beforehand. This would allow small fish through, holding back the adult fish, which were later 'harvested'. In addition the surrounding countryside would be left undisturbed to guarantee a supply of wild game at the time of the meeting.[15]

Gerritsen compared these fishing systems and the economy they supported with similar arrangements found on the rivers in the Pacific Northwest of North America. The people native to this region are credited with being among those who intensified food production and social complexity, thereby moving away from simple hunter-gatherer systems.[16]

The reluctance to credit engineered fisheries to colonised peoples, and thus underrate their sovereignty of the land, is not peculiar to Australia. American and Canadian researchers have known of the fish-trapping systems of the North American First Peoples for centuries, but only in the last decade was it discovered that Canadian First Nation peoples had been extending existing clam beds by building rock walls further away from the beach.

This enhancement of the existing resource is described by Judith Williams, one of the first to study this aspect of mariculture, which she describes as 'the bedrock of all husbandry, enhancing what is there'.[17]

The local First Nations people already knew of these clam 'gardens', and in some cases were still using them, but Williams was horrified to discover that the British Columbian Heritage Conservation Department had never inspected the hundreds of modified sites, and refused to do so after private archaeological analysis, simply because 'they were not in the literature' of First Nations' history.[18]

The archaeological authorities insisted that the Native Americans were at best complex hunter-gatherers and at worst 'wealthy scavengers', 'incapable of the sophisticated cultural development associated with agricultural societies'.[19] The prevailing orthodoxy was that 'hunter-gatherers were considered capable of only the most tentative claims to land ownership'.[20]

One further impediment to the revelation of this aspect of intensification was that the clam gardens were constructed and farmed by women and children — and such knowledge was never revealed to male archaeologists. Even after the engineering of the gardens had been examined by independent scientists, there was enormous reluctance to accept the results. The view seemed to be that such important structures could not have been overlooked by earlier archaeologists.

In Australia's case, however, such structures *are* in the literature in the form of explorer and settler journals and diaries. In *Australia and the Origins of Agriculture*, Gerritsen asks why we have been so slow in acknowledging this trajectory of Aboriginal and Torres Strait Islander people.

The knowledge to be gained by a more enthusiastic examination of Australia's past is not just an acknowledgement of Aboriginal prior ownership, but a search crucial to Australia's agricultural survival and the conservation of fish species.

One of the fish caught in the fishing systems of the Western District of Victoria was a galaxis, which the Aboriginal people called tuupuurn and which they had been farming for thousands of years.[21] Murray, when he arrived at Colac, stole a netfull of this little fish in his first hours in the country. He described its delicious flesh, but in the following years celebrated his fortune by rallying his neighbours to rid the earth of the original owners of the river system in which it bred. The fish has since disappeared as agricultural and industrial pollution of the lake system poisoned its habitat.

The Victorian Archaeological Survey examined stone arrangements at Lake Condah in the late 1990s, and declared that they couldn't be house sites. This was despite the vivid descriptions and drawings of such houses that had been made by early explorers and settlers in the region. A fellow archaeologist, Heather Builth, was critical of such a cursory investigation, and began an exhaustive study of her own.

The channels *looked* like man-made structures, but experts couldn't believe they would work hydrologically. The structures spread around the perimeter of the weirs *looked* like small round houses, but all sorts of speculation on their origins abounded, including conjecture that survivors of the fabled Mahogany Ship, a mysterious wreck seen near Port Fairy around the turn of the twentieth century, had built the systems. Local Aboriginal people were not credited with sufficient knowledge of engineering or energy to have created them. When the system flooded in the 1970s, researchers were amazed at how steadily water was introduced to all parts of the system and how, on its recession, the ponds were ideally suited to trap fish.

Builth, like Judith Williams in Canada, knew that only science could convince the doubters. She weighed and measured each stone in the house-like structures, analysed the results, and found that human agency was the only thing that could have produced such complexity.

So, if they were houses, and if the channels were a fishing system, then around 10,000 people lived a more or less sedentary life in this town.

Builth was conscious that similar structures were in evidence throughout the Western District. What was going on? If, she wondered, such a large population lived there, the demand for food would be extreme. There had to be some form of food preservation associated with this town.

She turned to the hollow trees she had noticed close to the workings, and could see immediately that they had all been used as fireplaces. An analysis of the soil at their base revealed eel fat. These timber tubes had been used as smoke houses for aeons, and Builth speculates on the huge quantities of smoked fish likely to have been processed in this way, and the likelihood that they formed the basis of trade with regions in New South Wales, South Australia, and other parts of Victoria.

Builth was guided in her research by the local Gundidjmara people, who knew all along what the structures represented, but whose opinion had never been sought. She also enlisted the aid of another scientist, Peter Kershaw, to attempt to find an age for the complex. Kershaw came to the conclusion that it was around 8,000 years old. This was when the area around Lake Condah had become inundated, presumably from the building of the channels, which had introduced water where it had never been before, thus completely changing the vegetation system around the banks.

The age of this aquaculture operation makes it one of the oldest in the world, preceding the period many scientists claim is the beginning of intensification in Australia.

The Victorian Archaeological Survey seemed to be restricted by their own assumptions of Aboriginal development, in the same way that so many pokers and prodders of Aboriginal culture seem not to have read the explorers' diaries. If they had, surely they would have gone further than the study of the kangaroo spear and digging stick in their analysis of Aboriginal economies.

In 1987, Frederick Rose, an experienced researcher of the Aboriginal economy, wrote *The Traditional Mode of*

Production of the Australian Aborigines, where he made a close analysis of Aboriginal economies. However, in his discussion of fishing he only looked at fish spears and shellfish collection. There was not one mention of traps or nets, despite the contact history being littered with such references.

It was not just in his writing about aquaculture that Rose seemed determined to place Aboriginal Australians in the hunter-gatherer basket. His study of housing is so limited that it looks deliberate, as nothing talks of permanence more directly than the house. Similarly, his study of flour production fails to mention the vast scale of grain harvests. He talks about the incredible labour required in milling and winnowing that grain, but never the scale of the operation — despite Sturt's description of the evening whirring of hundreds of mills grinding grain into flour.[22]

It is rare to come across a text after 1880 that describes the existence of intensive Aboriginal food production — either the fishing systems, or grain and tuber production. It's as if historians and researchers entered a dark room of enquiry, and firmly closed the door behind them. Thank God for people like Heather Builth and the very patient Gundidjmara. There are scales of Aboriginal production and degrees of sedentism, but it seems incredible for those who have studied the journals of our explorers and read their observations to insist that Australian Aboriginal people were nothing more than simple hunter-gatherers.

In his review of Gerritsen's *Australia and the Origins*

of Agriculture, Lourandos takes Gerritsen to task for depending too greatly on labels such as hunter-gatherer, complex hunter-gather, or agriculturalist. Lourandos argues that it is the process that is more important than the label.

The number of examples of Aboriginal harvest techniques, however, urges us to challenge the assumptions that under-rate Aboriginal agency. Aboriginal and Torres Strait Islander fishing systems were seen all over Australia in more or less complexity, and associated with more or less sedentary occupation. Anthropologist and zoologist Donald Thomson (1901–1970) remarked that the system of fishing nets and cane-fence structures on the north of Cape York and the brush fence he photographed were not just functional devices, but also things of beauty.

For example, Thomson photographed and described a complicated trap on the Glyde River. Behind a stout wall of stakes set into the riverbed, a cane platform had been erected. His photograph shows two large tubs made of paperbark and sewn with lawyer cane. Chutes of paperbark bring water through the fence and into the tubs. Men stand waist deep in the tubs facing the chutes in order to collect the fish as they are stranded on the bamboo platform. When I show this to students of Aboriginal Studies, they turn to me in astonishment, as if I'm pulling their leg. Australians make plaster figurines of Aboriginal men standing on one leg, spear in hand, waiting for the windfall kangaroo, while we have all but ignored ethnographic evidence of Aboriginal engineering.

Queensland fishing system

(D.F. Thomson)

Mitchell remarked on the huge scale of the fishing enterprises of the Paakantjyi on the Darling River, and, on the Bogan River, saw an intriguing method employed by women who fabricated a clever trap: 'A moveable dam of long, twisted dry grass through which water only can pass, is pushed from one end of the pond to the other, and all the fishes are necessarily captured.'[23]

The evidence of Mary Gilmore is downgraded by some academics, and one suspects that being a woman and a poet didn't help her cause.[24] The writing of Alice Duncan-Kemp is also often dismissed simply because she was a woman. Despite her having written a million words on Aboriginal life at Bidourie, the bush balladeer Barcroft

Boake is the only writer from south-west Queensland remembered in Australian literary history. It might seem a petty thing to mention, but eliminating Duncan-Kemp's first-hand observations compromises our understanding of the past.

Gilmore, however, was also recording things she'd seen in her lifetime, and things she remembered being told by her family. Her remembrances are clear and detailed, and cover numerous methods and locations across New South Wales and southern Queensland.

Early surveyor Mortimer William Lewis saw fishing complexes on the Warburton Creek, Northern Territory, of such permanence that the tribes were sedentary, while Davis and McKinlay in their search for Burke and Wills saw the abundance of fish being harvested in the Strzelecki Desert.[25] Sturt saw weirs and sluices on almost every flowing river. Permanent fisheries were an established part of the Aboriginal economy from one side of the country to the other.

Thomson's record of canoe-making for fishing and birding in Arnhem Land's Arafura Swamp has been celebrated in the film *Ten Canoes*, but large, organised fishing expeditions by watercraft were observed all around the coast.

Paul Memmott, an anthropologist and architect, records that in 1995 over 334 individual fish traps could be seen on Wellesley Island in the Gulf of Carpentaria, and nearby on the mainland coast.[26] The Kaiadilt clan lived on the produce from these traps, and named four types of shark and two types of stingray they caught in the traps, along with innumerable smaller species.

Wellesley Island fish traps
(Connah and Jones)

Abalone

The Australian coast, from just south of Perth, across the southern coast, including Tasmania and her islands, west to Gippsland in eastern Victoria and north to Wollongong in New South Wales, supported the Aboriginal abalone harvest.

The skeletons of women from Victoria's coastal regions were found to have an odd bone-growth in the ear. Scientists recognised that the bone had thickened to protect the ear from extreme cold; the women were diving for abalone. They had what doctors refer to as 'surfers' ear'.

Diving for crayfish and abalone was an important part of the southern coastal economy. Not all the diving was done by women, but in Tasmania and Victoria the shellfish was collected predominantly by women.

Uncle Banjo Clarke, a Keeraywoorrong elder, described to me the process by which men netted and speared crayfish. The most intriguing process was how men would swim to a reef and hang onto the kelp while feeling with their feet for the feelers of crayfish. They would then dive down and grab the feelers, and haul the crayfish from its cave.

Access to large fish resources was common in Australia, and the methods employed to harvest them varied. Some resources were so productive that they allowed many communities to live a sedentary or semi-sedentary life close to their fish traps or their fishing grounds.

Women from the Eden area in southern New South Wales have written a book, *Mutton Fish*, chronicling the Yuin and neighbouring tribes' reliance on the abalone economy.[27] The shellfish was a favoured high-protein item of the coastal diet in this part of Australia.

The shell of the abalone is beautifully coloured rainbow pearl. It was used in the making of traditional jewellery, but breaks down very quickly and is under-represented in living site remains as a result. The shells of turbo (warrener) or limpets (bimbula) are much more prevalent, although they probably represent a fraction of the protein yield.

Archaeologists can only measure what they find, so soft-skeleton and friable-shell creatures such as crayfish, whiting, shark, abalone, urchin, and snapper are often under-represented in surveys of the Aboriginal maritime economy.

European settlers shunned the abalone, and referred to it derogatorily as mutton fish. English cooking has never enjoyed much of a reputation, and so the colonial chefs applied their most subtle cuisine to the abalone: boiling. The flesh of the abalone takes on the texture of industrial rubber when handled this way, and so, with no other cooking refinement to fall back on, the 'mutton fish' was considered food for 'blacks'. The Chinese and Japanese knew otherwise. They pounded the fish, sliced it finely, and cooked it quickly — for no more than thirty seconds — in a searing pan. Treated this way, the flesh is delicate, tender, and full of flavour.

One Aboriginal recipe suggests cooking the abalone in its own shell on hot coals. I tried this, expecting the flesh to toughen under these conditions, but instead found that it remained tender and even more flavourful.

Once entrepreneurs realised that Australian Chinese were exporting abalone meat, they lobbied the Departments of Primary Industry to establish licensing, quotas, and closed marketing boards, which operated like cartels.

Aboriginals are now seen as poachers simply because the shellfish is so enormously valuable. When it was 'mutton fish', they were allowed to harvest as much as they wanted. Today they are jailed for pursuing their traditional harvest.

Watercraft

When one of the early sailing ships limped around Victoria's Cape Otway after a vicious storm in Bass Strait, the captain was astounded to find numerous canoes fishing in benign waters in the lee of the Cape. All the canoes were sailed by Gadabanud women.

The occupation of Montague Island, nine kilometres off the coast of New South Wales at Narooma, had to be negotiated by canoe, as sea levels were never sufficiently low to make it accessible by land. One of the enduring stories told by the local people is of a terrible calamity when a fleet of canoes was overwhelmed by a sudden squall as they returned from the island.

Lady Julia Percy Island, ten kilometres off the Victorian coast near Portland, was also inaccessible, even during low sea-level periods. However, it was occupied extensively by Aboriginal people, who called it Deen Maar.

Watercraft were a significant tool in the Aboriginal and Torres Strait Island fishing economy. Rottnest

Island, eighteen kilometres off the Western Australian coast from Fremantle, could be reached by land around 12,000 years ago. However, some of the artefacts currently under examination by archaeologists are reputed to be 60,000 years old. Ocean voyages must have been undertaken to reach the island under the lure of the enormous seafood resources.

In arguably his most evocative writing, Robinson compared the night fleets fishing on the Murray to the harbour at Teneriffe, while Beveridge said of the fleets he observed, '[T]he flotilla presented a scene so quaintly striking as to be well worthy of an artist's pencil'.[28]

Canoes with outriggers were built in the north of the country and in the Torres Strait Islands, and were seen well out to sea hunting for pelagic fish.[29] Small sails were also used for ocean canoes when line and net fishing. On inland waterways, canoes were employed as punts to cross rivers, and were often described by explorers as the only way they could continue their adventures of discovery. Many early photographs of Australian waterways feature various canoe designs and the nets and traps associated with them.

Yuin people of the New South Wales south coast have re-introduced the canoe-making tradition; several fine examples have been produced, including one made on the banks of the Brogo River.

Yuin Gurandgi with a canoe made at the Brogo River in 2015
(Stephen Mitchell)

When I visited members of the Pascoe family at Lockhart River on the Cape York Peninsula in 2009, I was struck by the number of people involved in the fishing industry. Every backyard seemed to have a tinny. Those east coast Cape communities have vegetable and fruit products in the jungles and on the plains, and harvest them still, but their eyes are forever turned to the sea.

Stephanie Anderson translated a terrific book about the French sailor Pelletier, who was cast away on this part of the coast in 1858 and lived for some time with the local Aboriginals. Hints of early reliance on the produce from the sea can be found in her account. Pelletier also discusses the people's management of yam production to ensure a supply during the dry season, but it is the sea from which those people gained most of their nutrition and cultural solace.

The early history of Australia is crowded with references to Aboriginal watercraft and fishing techniques, yet Australians remain strangely impervious to that knowledge and to the Aboriginal economy in general.

You can read other theories of Aboriginal culture, spirituality, and economy in New Age texts, or the books of over-enthusiastic researchers, but often they make guesses to bridge the gaps in knowledge. Too often, they ascribe all sorts of mystical wisdom to their subjects, but their earnest romanticism is unnecessary, as the observations of the first explorers and settlers provide an enormous body of material. In this book, I am drawing on only a small sample of what is available to any Australian with a computer mouse or a library card. The reason I have provided so many examples, however, is to emphasise the depth of the available material and the desperate need for a revision of our history.

3

Population and Housing

Collecting such a welter of evidence might seem a tedious excess to some, but reference to Aboriginal housing is so remote from the Australian consciousness that, on reading of one or two examples, people might be encouraged to see them as aberrations. I have used examples from across the continent in order to show the pervasive nature of these cultural developments.

Houses and villages were observed from the far Kimberley to Cape York, from Hutt River to Tasmania, from Brewarrina to Hamilton. Permanent housing was a feature of the pre-contact Aboriginal economy, and marked the movement towards agricultural reliance. These villages were not just functional occupation centres, but places of solace and comfort in often difficult terrains and climates.

New examples are being discovered all the time. Archaeologists are currently examining a complex village site in 'Australia's dead heart' where the people had a complex water-management system, sophisticated housing, stone quarries, and seed-grinding and storage arrangements. This is a major cultural site where people employed engineering to manage the environment. It has the potential, together with the examination of the hundreds of similar sites around the country, to provide a very different picture of Australia's social, economic, and cultural history. Several of these villages have proven to be the oldest in the world, a discovery that suggests Aboriginal people also invented society. Genuine interest in the Australian past will provide a rich vein of knowledge that will inform our attempts to live in this country but still protect the environment.

Sturt's saviours

When Charles Sturt arrived in Australia, the interior was still largely unknown to Europeans. Efforts to penetrate to the centre of the continent had been thwarted by the terrain and inhospitable conditions.

Sturt's expedition, beginning in 1844, was hampered by the rigours of the environment. It was so hot that thermometers burst, screws fell out of boxes, and the lead from pencils.

Sturt's party reached Cooper's Creek, in what was

to become known as Sturt's Stony Desert, where they were confronted by sand ridges thirty metres high. They ploughed on, enduring incredible hardships. Sturt climbed one final dune and peered down onto the plain. His journal records:

> [O]n gaining the summit [we] were hailed with a deafening shout by 3 or 400 natives, who had assembled on the flat below ... I had never before come so suddenly upon so large a party. The scene was of the most animated description, and was rendered still more striking from the circumference of the native huts, at which there were a number of women and children, occupying the whole crest of a long piece of rising ground at the opposite side of the flat.[1]

Sturt was looking on the dry floodplain of a river, and he couldn't understand how these people were able to survive. Sick and weary, and with horses stumbling with hunger, thirst, and fatigue, Sturt was alarmed at coming so suddenly on so many Aboriginals:

> Had these people been of an unfriendly temper, we could not in any possibility have escaped them, for our horses could not have broken into a canter to save our lives or their own. We were therefore wholly in their power ... but, so far from exhibiting any unkind feeling, they treated us with genuine hospitality, and we might

> certainly have commanded whatever they had. Several of them brought us large troughs of water, and when we had taken a little, held them up for our horses to drink; an instance of nerve that is very remarkable, for I am quite sure that no white man (having never seen or heard of a horse before, and with the natural apprehension the first sight of such an animal would create) would deliberately have walked up to what must have appeared to them most formidable brutes, and placing the troughs they carried against their breast, they allowed the horses to drink, with their noses almost touching them. They likewise offered us some roasted ducks, and some cake. When we walked over to their camp, they pointed to a large new hut, and told us we could sleep there … and (later) they brought a quantity of sticks for us to make a fire, wood being extremely scarce.[2]

Sturt was doing it tough among the savages alright. New house, roast duck, and cake!

One reader of the first edition of *Dark Emu* (2014) thought I was inventing material in order to fabricate Aboriginal achievement, but Sturt's journal is quite definite about his admiration for Aboriginal people — although he had a low opinion of the beauty of Aboriginal women. The opinion of European men in regard to Indigenous women, however, was not without its prejudice or darker motivations.

Peter Gebhardt, poet, and Michael Perry, engineer, volunteered to read the journals with a view to examining the roast duck-and-cake story. Part of that test was to sort out Sturt's longitudes and latitudes, because he didn't record them every day. We are confident, however, that the incident described above occurred on 3–4 November 1845, just north-east of today's Innamincka at a latitude of 127 degrees 47 minutes south and a longitude of 141 degrees 51 minutes east.

Sturt saw buildings in a number of different locations, including several at Strzelecki Creek in the same year, and he made sketches of many. The entrance of one was 14.5 metres wide and two metres high, the roof having been plastered with a thick coating of clay.

In another area, he saw huts made with strong elliptical arches, covered with boughs, and rendered with 'a thick coating of clay so that the huts were impervious to wind and heat. These huts were a considerable size, and close to each there was a smaller one equally well made ... and had apparently been swept prior to the departure of the inhabitants.'[3]

Later, on the same journey, he remarked, '[T]he paths of the natives became wider and wider as we advanced. They were now as broad as a footpath in England, by a roadside, and were well trodden; numerous huts of boughs also lined the creek, so that it was evident we were advancing into a well peopled country.'[4]

He added, 'Where there were villages these huts were

built in rows, the front of one hut being at the back of the other, and it appeared to be a singular but universal custom to erect a smaller hut at no great distance from the large ones.'[5]

Sturt expressed his surprise and admiration for many elements of Aboriginal ingenuity, construction, and food production, but his main focus remained the occupation of the fertile areas. His brother Evelyn was already involved in overlanding cattle into the areas that Sturt had discovered.

Despite the very pragmatic nature of his ambitions, it is a telling indication of Sturt's respect for Aboriginal achievement that he mentions them at all in his journals. He was not an apologist or a bleeding heart, and he occupied over 5,000 acres himself, but the evidence of the scale and sophistication of the buildings forced him to recognise their significance.

The timber frames of buildings similar to those seen by Sturt were still standing in the 1970s, despite their having been built perhaps sixty years earlier. The remoteness of these houses preserved them from destruction by cattle or fire, and allowed researchers to witness their size and permanence.

In 1861, George Goyder, an early South Australian surveyor-general, travelled through many of the districts explored by Sturt, and encountered many large buildings. One was 'a large structure … in a "settlement" south west of Lake Blanche in South Australia … constructed in a similar manner to those described by Captain Sturt, and very warm and comfortable, the largest capable of holding thirty to forty people'.[6]

Both Karl Emil Jung and Howitt reported large cone-shaped huts made of 'heavy wooden logs' on Cooper's Creek east of Lake Eyre. Thomas Mitchell discovered a number of large dwellings near White Lake, close to Victoria's Grampian Ranges, and residents of Lake Condah told the Rev. Francis that prior to the arrival of Europeans they had lived in communities of forty or more, and all living in the same hut.[7]

Houses in the Illawarra region of New South Wales 'were often quite elaborate, similar in size and form to the tepee of the American Indian, though made entirely of local wood and plant fibre'.[8]

Even in the harsh and difficult terrain of the sand dunes of the desert, Sturt encountered Aboriginal wells and substantial houses. Several villages were located near Birdsville, south-west Queensland, where today the remoteness and inhospitable nature of the land is mythologised as the desolate Outback. Many Australians find it hard to imagine the area as a once productive and healthy environment for large numbers of Aboriginal people.

One of the great Australian ironies is that the desert in which Sturt nearly died twice was peopled by large numbers of Aboriginal people taking advantage of plains to manage flood dispersal, plant crops, and harvest and store grain. Any who passed through the frontier noticed similar populations. The Australian poet Mary Gilmore reported her uncles as saying that 5,000 people lived around the Brewarrina fish traps. Townships of 1,000 or more people were seen in Western Victoria. Sturt himself saw a prosperous town of 1,000 on the banks of the Darling River, and Duncan-Kemp claimed that 3,000 lived on Farrar's Lagoon, not far from Brewarrina.[9]

It is amazing that such witness is not part of Australian geographical and historical folklore. Such is the tenacity of the Australian delusion, it encourages an impoverished national debate. In September 2010, the Democratic Labor Party's Legislative Council member, Peter Kavanagh, in decrying a State of Victoria Native Title Bill, argued that nomads had no concept of land ownership. He argued that Native Title would 'entrench a system of racial discrimination in favour of one group against every other group. That is, against non-Aboriginal Australians'.[10]

This stain is deep in our chalk, and until we can accept what the explorers saw as part of the national story, our debate about national origins, character, and attributes is hobbled by ignorance. That ignorance began with the first Europeans to visit the country. Even men as enlightened as Mitchell and Sturt, despite having seen villages of over a thousand people, and grain fields reaching to the horizon, lapse into imperial euphemism by referring to those same people as 'children of the soil', 'sable friends', and 'knights of the desert'.[11] On seeing houses built to accommodate forty people in groups of fifty or more, both explorers resort to words such as 'huts' or 'hovels' to describe buildings that in rural Ireland would have been called croft houses.

The determination of the colonial governors, surveyors, and explorers to discount Aboriginal achievement has persisted into contemporary Australian society. When interviewing the authors of the *Encyclopaedia of Australian Architecture* in November 2011, Radio National presenter Alan Saunders expressed surprise that the first chapter was devoted to Aboriginal architecture. He wondered at the depth of Australian scholarship and education that had allowed Australians to remain ignorant of this aspect of our national culture.

Any country will have naysayers among its citizenry, be it regarding climate change, birth control, taxation, gun control, or speed limits; however, if the general population persists in hiding from obvious facts of history, we are destined to repeat the selective opinions of the colonists. There are researchers querying the accepted wisdoms, but it is up to our politicians and educators, and ultimately the public, to reconsider the pre-contact Aboriginal economy.

Large populations of Aboriginal people were manipulating the Australian environment and husbanding plants to produce surplus food of such great quantity that populations could lead more or less sedentary lives in the vicinity of their crops. Well-trodden roads attested to the size of the populations observed by Sturt, Grey, and others, as did wells and irrigation dams; and, of course, houses grouped in villages spoke for their industry.

New examples of sedentary or semi-sedentary Aboriginal culture are being discovered all the time. Michael Westaway is currently examining a complex village site in northern Queensland. These patterns must be accepted as part of our historical narrative.

Of course, not every Aboriginal clan was engaged in agricultural production to the same extent, and not all lived in the houses the explorers describe. I have already cited examples of large-scale plantings from areas in a wide range of climatic zones, from which it is obvious that this was a central part of the Aboriginal lifestyle.

It's not just the size of the population or the scale and number of the houses that needs to be considered, but also their strength, aesthetics, and comfort. Design features had been developed to make harsh environments habitable.

On an early expedition, Sturt had seen a sophisticated village of seventy domed huts on the Darling River, each capable of housing up to fifteen people:

> [The houses] were made of strong boughs fixed in a circle in the ground, so as to meet in a common centre; on these there was ... a thick seam of grass and leaves and over this a compact coating of clay. They were from eight to ten feet in diameter, and about four and a half feet high, the opening into them not being larger than

> to allow a man to creep in. These huts also faced north-west, and each one had a smaller one attached.'[12]

Some have speculated that the low doorway and the smoky internal fires discouraged flies from entering. Paul Memmott wondered if the low light-levels in buildings with small doorways also acted as a fly deterrent. Most dome houses also used small bark, timber, or matting doors to eliminate insects.[13]

Many early observers commented on the aesthetic proportions, tasteful positioning, and social harmony of the townships. Sturt described one town as evening fell:

> [T]he natives ... sat up to a late hour at their own camp, the women being employed beating the seed for cakes, between two stones, and the noise they made was exactly like the working of a loom factory. The whole encampment, with the long line of fires, looked exceedingly pretty, and the dusky figures of the natives standing by them, or moving from one hut to the other, had the effect of a fine scene in a play. At eleven all was still, and you would not have known that you were in such close contiguity to so large an assemblage of people.[14]

The smaller huts attached to the dwellings of this and other clans were full of stored produce. Yards attached to these store houses were used as animal holding pens. People here were not clinging on to survival in the desert;

they were thriving, and engaged in a rich and joyful life.

They made provision for their comfort by allocating extensive resources to the building of their homes and gardens. Gammage refers to wells in north-west Victoria up to two metres deep, and where bushes had been deliberately 'twined' above the well to provide an elegant arbour and shade protection for the water source.[15] He also quotes Giles, who found several large wells, one with a bank almost two metres high and twenty metres wide.[16] Another South Australian well was three metres deep, and had a shaft at its foot driven at right angles to tap a spring. Holes had been cut into the sides of the shaft to provide access. About these areas, glades and grasslands had been cultivated after water had been assured by industry and innovation.[17]

Mitchell's revelation

Thomas Mitchell admired another large village on the Gwydir River, in the Murray Darling basin:

> In crossing one hollow we passed among the huts of a native tribe. They were tastefully distributed amongst drooping acacias and casuarinae; some resembled bowers under yellow fragrant mimosae; some were isolated under the deeper shades of casuarinae; while others were placed more socially, three or four together, fronting to one and the same hearth. Each hut was semicircular, or circular, the roof conical, and from one

> side a flat roof stood forward like a portico, supported by two sticks. Most of them were close to the trunk of a tree, and they were covered, not as in many other parts, by sheets of bark, but with a variety of materials, such as reeds, grass and boughs. The interior of each looked clean, and to us passing in the rain, gave some idea, not only of shelter, but even of comfort and happiness. They afforded a favourable specimen of the taste of the gins, whose business it is to construct the huts.[18]

In the region of the Darling River, Mitchell saw another small town where each hut could accommodate up to fifteen people, being substantial constructions with thatches over 30 centimetres thick. 'These permanent huts seemed also to indicate a race of more peaceable and settled habits.'[19]

Mitchell became a great admirer of the Aboriginal house, and wrote: 'I began to learn that such huts, with a good fire before them, made very comfortable quarters.'[20] Mitchell was sensitive to the quality of the houses, but insensitive to his occupation of someone else's residence. He occupied empty houses on many occasions, and liberties of this kind were likely to have ruptured the relationship between white and black more severely than any action other than physical attack.

We are, however, in Mitchell's debt for the detail and eloquence of many of his observations, and I can imagine having a long and pleasant yarn with this empathetic and intelligent man. But even though he relished and appreciated his witness of this incredible civilisation, it brought no halt to his search for grasslands. Despite his admiration of the housing structures, and of the industry and innovation required to produce them, he reserved his greatest praise for the land, and the wealth it would afford the conqueror.

Mitchell's expressions of concern for the Indigenes might warm our heart, as many historians still feel warmed by La Trobe's civility, but Mitchell wrote in the very next sentence, 'If wild cattle on the contrary become numerous the natives might also increase in number and, if not civilised and instructed now, might become formidable and implacable enemies then, as no absolute right to kill even wild cattle would be conceded to them.'[21]

We like to extol the virtues of our gentlest and most refined forebears, often ignoring the murders committed by their thugs. However, a closer reading of what our ancestors wrote about Aboriginal society, and how much of their time it occupied, might cause us to reassess their generosity. Mitchell's mind, for instance, is often distracted by the urbane ambitions of town planning and civil engineering, when a scientific analysis might have been revelatory:

> The growth of a town depends very much on the direction of great roads, and must be more certain, and the allotments more valuable, when the most eligible

> line of thoroughfare is ascertained ... Such works of public convenience should precede, as much as possible, the progress of colonization. The plan ... should be well considered, before the capital ... is applied ... The convenience of the public, and the encouragement of the mechanic, who is indeed the pioneer of colonists cannot be sufficiently studied, in affording facilities for the establishment of inns, and the growth of population along great roads.[22]

Later he writes, in bucolic prose:

> Peace and plenty now smile on the banks of 'Wambool' and the British enterprise and industry may produce in time, a similar change on the desolate banks of the Nammoy, Gwydir, and Karaula, and throughout those extensive regions beyond the Coast range, still further northward, — all as yet unpeopled, save by the wandering aborigines, who may then, as at Bathurst now, enjoy that security and protection, to which they have so just a claim.[23]

Even in 1835, Mitchell must have been aware that peace and security was never brought to the dispossessed on any continent. He is a good, industrious, and optimistic Scotsman, but his prejudice hides from him the fact that he is a crucial agent in the complete destruction of Aboriginal society. He is so full of visions of inns and roads and romantically smoking chimneys that he ignores things which are staring him in the face.

We have become used to thinking of Mitchell as being fascinated by Aboriginal life, but in fact this topic commanded very little of his attention. He passed tilled land frequently, and often blessed the discovery with less than a line or two. He found intricate fish traps, but barely paused to describe them. On the journey out to chart the Darling River in 1835, his writing was full of speculation about the position of roads and great estates that might take advantage of the open grasslands created by the Aboriginal people.

Even though he was in a position to analyse Aboriginal land use, he often seemed confounded by the order and beauty of the parklands he crossed, and resorted to describing the artistry of 'careless nature'.[24]

Anthropologist Tony Barta comments that Mitchell's 'portrayal of Aborigines and their cultural attainments is notably positive. But they do not figure in his vision.'[25] There's no doubt that Mitchell was much more sensitive to Aboriginal attainment than most of his contemporaries. Nonetheless, in reading the introduction to his *Three Expeditions into the Interior of Eastern Australia*, readers will be struck by the paucity of his references to Aboriginal people, and by the casual dismissal of their rights on those occasions when he does mention them.

Mitchell's party killed seven Aboriginals near Mount Dispersion in 1836, an unknown number between

Wilcannia and Menindee in 1835, and one at Lake Boga. He wrote that overlanding parties would need to be large and well armed to ward off Aboriginal attack.[26]

In a letter to Governor Bourke describing the Mount Dispersion attack, which he argued had been necessary to deter the possibility of Aboriginal aggressions, he wrote: 'My men pursuing and shooting as many as they could — numbers were shot in swimming across the Murray.' Bourke ordered that the reference be deleted from the *Gazette*.[27]

Editing references to violence committed against Aboriginal people and the evidence of their established villages and economy was not uncommon. Gerritsen comments:

> The suppression, or discouragement of public disclosure of permanent settlements and more sedentary existence may have been yet another factor contributing to a distortion in historical information about relevant groups, and hence modern understandings. For example, in his original account of his brief reconnaissance in the Victoria District in early December 1841, published at the time in a Perth newspaper, Captain Stokes of the Beagle referred to 'their [the Nhanda's] winter habitations substantially constructed'. But that very line is omitted, the only one to be so, when the substantive part of the newspaper account was reproduced in his published journals in 1846. Similarly, in Victoria, a Select Committee of the Legislative Council in 1858–59 took

> evidence from Thomas and Sievewright (Aboriginal Protectors) in which both discounted that there had been any permanency of settlement, even though we know they knew differently. The suspicion is that there was intent to discredit evidence of permanent settlements because of the implications this may have had for the morality and legality of the colonial dispossession.[28]

Reading Mitchell's journals, and admiring his verses and sketches, it is impossible not to notice how he differed from most of his contemporaries. Of the verdant plains he found in his Australia Felix of western Victoria, he wrote, 'there is not in the wide world a valley so sweet', and later, 'Certainly a land more favourable for colonization could not be found.'[29] You can imagine his regret at the murders and attacks, but it did not stay his hand; he did not resile from murder, if resistance were to interrupt his expeditionary force.

I can't think of any explorer who gave up his excursions for fear of having to kill those who resisted his advance, or any settler who desisted from taking up ground because it would be of disadvantage to the Indigenes. Even James Dawson, the Aboriginals' greatest friend in Victoria's Western District, still took possession of land that he knew had formerly belonged to his friends.

Settlers and explorers were united in their assumption of superiority and entitlement. The hoofs of the horses and the tramp of their feet never paused, despite their knowledge that they were displacing a sophisticated society with a complex economy. They must have been acutely aware of the permanence and prosperity of many regions, because the early literature has so many examples of their witness.

The academic Alan Pope refers to this as 'the incompatibility of establishing individual fortunes based on ownership and exploitation of land and the maintenance of an Aboriginal way of life so fundamentally linked to that same land'.[30]

On the Darling River, explorers saw similar towns to those seen by Sturt and Mitchell, and estimated the population of each to be no less than a thousand. Peter Dargin estimated the population of the region as 3,000, but the journals of Sturt, Mitchell, and others reveal that they passed many such populous villages. These figures strongly contradict both current and past assumptions of a sparsely populated pre-colonial land.

Government surveyor David Lindsay reported many large villages at Poeppel Corner (a corner of state boundaries in Australia, where the state of Queensland meets South Australia and the Northern Territory), including one house large enough to accommodate thirty to forty people.[31] Mitchell saw villages comprising similar houses on the Barcoo River in Western Queensland, but added that some had several rooms each. The villages were associated with extensive networks of well-worn paths, many of which have become today's highways.

The early pastoralist John Conrick wrote in his papers of a house over ninety feet (thirty metres) in circumference, which was used for holding corroborees. Goyder described other buildings, similar in construction to those reported by Sturt, as 'very warm and comfortable', with the largest having the capacity to hold thirty to forty people.[32]

John McDouall Stuart, who hardly mentions Aboriginal people in his journals of exploration, found beehive huts made of mud, and remarks on the beauty and comfort they afford.[33]

In 1883, David Lindsay reported from his survey of Arnhem Land that he:

> came on the site of a large native encampment, quite a quarter of a mile across. Framework of several large humpies one having been 12 feet high: small enclosures as if some small game had been yarded and kept alive … This camp must have contained quite 500 natives, and have been the site of some great festival, the corroboree or dancing grounds, being numerous and well worn.[34]

But in describing what must have been large buildings, he resorted to the word 'humpy'.

Even Edward Curr, the old curmudgeon of the Murray River, admitted that the sophisticated bark huts made by Aboriginal people were easily the most comfortable of any habitations in the colonial bush. At Mallacoota in 1842,

Joseph Lingard met two Aboriginal men and 'made bold to go into their retreat, which I found to be like a house inside'.[35]

Gerritsen lists numerous reports by explorers and early settlers who saw large villages. Mitchell referred to the banks of the Darling, where 'the buzz of population gave to the banks at this place the cheerful character of a village in a populous country'.[36] As Gerritsen summarises:

> Most Australians would find such comments rather surprising given that much of this area is usually seen as being desolate and inhospitable. They even feature in some well-known expressions such as 'back o'Bourke' or 'the never never', forbidding places where no person in their right mind would go.[37]

But people did live there and prospered; towns thrived because the inhabitants were utilising the natural conditions and developing the endemic grains and tubers. Unfortunately, the destruction of both grasslands and villages was swift.

Elizabeth Williams quotes a graphic account from William Thomas, an Aboriginal Protector, which provides a neat summary of the scale and sophistication of Aboriginal housing, but also why so few people saw it after the first visits of Europeans:

> [The] first settlers found a regular aboriginal settlement. This settlement was about 50 miles NE of Port Fairy.

> There was on the banks of the creek between 20 and 30 huts of the form of a beehive, some of them capable of holding a dozen people. These huts were about 6' high or (a) little more, about 10' in diameter, an opening about 3'6" high for a door which was closed at night if they required with a sheet of bark, an aperture at the top 8" or 9" to let out the smoke which in wet weather they covered with a sod. These buildings were all made of a circular form, closely worked and then covered with mud, they would bear the weight of a man on them without injury. These blacks made various well-constructed dams in the creek which by certain heights acted as sluice gates at the flooding season … In 1840 a sheep station was formed on the opposite banks of the creek …. and one day while the blacks were away from their village up the creek, seeking their daily fare, the white people set fire to and demolished the aboriginal settlement … What became of the blacks (my informant) would not tell but at the close of 1841 … he could not trace a single hut along the whole creek.[38]

Construction and design

Despite the early destruction of these dwellings, there was still evidence of this old way of life elsewhere. Examples of Aboriginal peoples' buildings were being described into the twentieth century. Aboriginal people in the southern Gulf of Carpentaria area created large, domed, grass-covered shelters with small entrances so that the opening could be easily closed. These structures were adapted so the wet season could be survived in comfort, and so insects could be repelled by having a small, smoky fire within.[39]

Other peoples in the Gulf region and Torres Strait Islands built complicated structures on stilts or beautifully domed buildings constructed of great arcing bamboo canes covered in dense grass thatching. These buildings catered for large extended families.[40]

The Alyawarr people near Tennant Creek had less complex housing as a result of more benign conditions, but included in the encampment a small domed structure and a yard for their dogs.[41]

The people of Cape York and Arnhem Land, where the seasons were divided into the wet and dry, usually had two seasonal camps and two different styles of housing.[42] There were large, thatched, waterproofed and domed wet season huts, and in the dry they used lighter, more airy buildings. Wet-season occupation of these difficult areas was made possible by the storing of starches from pandanus fruits within secure buildings.

The use of the poisonous pandanus surprised many early European visitors until it was learnt the alkaloids were removed by a leaching and pulping process. The foods were pounded, milled, and roasted before storage in the huts for the wet season, when access to the rainforest was limited. The leaves were also crucial components of thatching and weaving.

The rainforest peoples had an array of language names for each of the housing types they used, including the larger buildings, which could accommodate thirty or more people.

On the Mitchell River, Queensland, clans employed complex intersecting dome structures clad in paperbark and palm leaves. Western coast Tasmanians also built domed, waterproof structures to insulate them from the wet and cold.

The mud-coated domes were reported by numerous witnesses, including explorer Ernest Favenc, whose treatment of and regard for Aboriginal people was appalling. In 1877, he travelled through Central Australia to survey a route for the overland telegraph, where he saw many Aboriginal towns — but even in reporting on large and sophisticated structures, he managed to denigrate them with the term 'hovel'.

The floors of many of the domed structures of inland Australia were substantially lower than the surrounding ground, presumably for the purpose of retaining heat during the cold desert evenings.

Stone was sometimes used as an alternative to clay daubing, the interstices between stones being mortared with mud. The framework of such buildings was necessarily quite substantial. Occasionally, domed dwellings had a small veranda attached over the doorway, with a single wall on the weather side to provide protection for a fire lit in the doorway and the comfort of anyone sitting outside the house.

The dome style was also used by large family groups as a shade house during the day. Once again, the framing was of substantial interlocking ribs covered in foliage and grass. John Oxley described similar houses near Moreton Bay in Queensland as being 'quite impervious to the rain thus forming a spacious and commodious hut, capable of containing from ten to twelve people'.[43]

Many of the above examples come from Paul Memmott's excellent book, *Gunyah, Goondie and Wurley: the Aboriginal architecture of Australia* (2007), in which he uses the work of early photographers such as Donald Thomson, who were able to visit remote communities where traditional economies were still being practised.

Memmott noted that some areas within Arnhem Land developed an array of housing styles: sheets of paperbark over rectangular structures supported by forked posts and with internal raised sleeping platforms; tent-like structures with bark over a ridge pole; shade houses of various designs; platforms built in trees in swampy areas; and a modification of the domed structure for the wet season, which employed overlapping sheets of paperbark.

The domes were constructed with rods of flexible saplings criss-crossing each other in similar fashion to modern flexible tent poles, the difference being that the Arnhem Land model had many more poles. The entrance was reinforced with paperbark and cord, so that those leaving and entering wouldn't damage the structure. The door was only small enough to crawl through, so that it could be easily sealed against rain and mosquitoes.

Pointed dome house
(Queensland Museum)

We are indebted to Thomson's excellent photographs collected during his many years living with the people of Arnhem Land, because so few others thought to photograph or draw Aboriginal housing.

Thomson reported that fan palms were often used as a cladding, and produced a very aesthetically pleasing result. Woven cladding was sometimes preferred, with an equally pleasing appearance. This type of cladding was also seen in the Kimberley.

Building types varied according to the materials available. Clay from lake shores or ant-nest material are both excellent building materials. Where safe cave systems occurred, they were also used for housing and ceremonial purposes. In one dramatic example filmed for the ABC TV series *First Footprints,* Arnhem Land people had removed large amounts of stone and soil, leaving a massive columned gallery to exhibit extraordinary art. Viewers reacted in astonishment after seeing this construction, and flooded the ABC with questions and comments. This was a dramatic piece of cultural evidence, and yet it was seen by many Australians for the first time in 2013.

Similarly, many Australians were surprised when Aboriginal women were featured at the 2006 Melbourne Commonwealth Games opening ceremony wearing magnificent possum-skin cloaks. The pervasive idea is that Aboriginal people wore either nothing or animal skins.

And they did wear skins, but they were sewn, had sleeves for the accommodation of infant children, and could be used as rugs and bedding. Thankfully, the art of the manufacture is being revived by a group of Victorian Aboriginal women, Vicki Couzens, Lee Daroch, and Treanna Hamm, in a hugely important cultural resurgence.

When Heather Le Griffon was researching her book, *Campfires at the Cross*, she remembers seeing her first skin cloak in the Victorian museum: 'I was shown a rare example of a possum skin cloak. I had expected to see rough overcast stitches cobbling the skins together. When I saw the beautiful stitches sewn by the needlewoman, I was moved to tears, shamed by the poverty of my expectations.'[44]

The cloaks were made after pegging out and drying the skins, and later scraping and incising them to render the skins more flexible. The incisions were part of intricate designs, and the cloak was sewn together with finely crafted bone needles and thread made from the dried sinews of kangaroo tails. They were works of art, and consummately crafted.

If a woman as earnest, intelligent, and well versed in Aboriginal history as Le Griffon could have her prejudices exposed so simply, all of us must be alert to that greatest of all limitations to wisdom: the assumption.

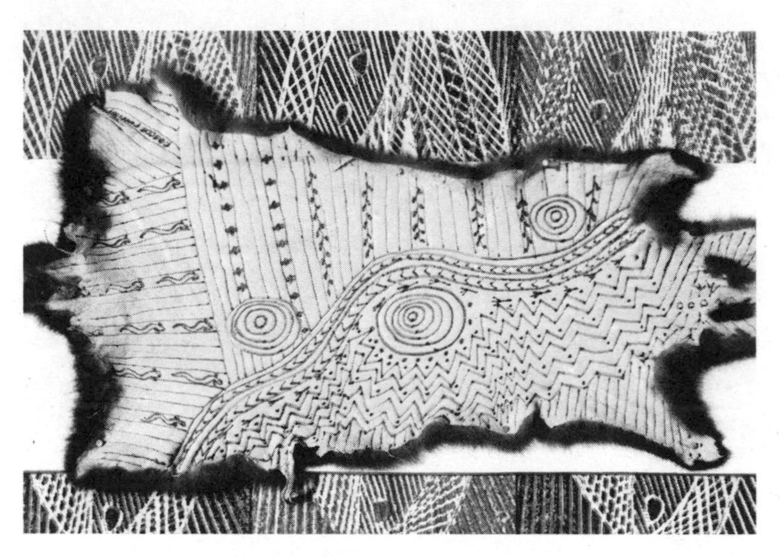

Possum-skin design by Bruce Pascoe

(Lyn Harwood)

The manufacture of cloaks, hats, shoes, and skirts is a study on its own, but, once again, the evidence of that industry, as exposed by explorer and settler diaries, is waiting for Australians to fully appreciate Aboriginal achievements.

Stone structures

Reviewing the earliest colonial reports reveals an astonishing amount of reference to stone houses and other structures. The first explorers and settlers must have seen many more houses than the ones they wrote about; but even with such reluctant witnesses, the evidence is overwhelming.

The remains of many stone houses can still be seen today. Lourandos described them at Cape Otway and Builth at Lake Condah, and research continues in the Blue Mountains in New South Wales, High Cliffy Island, Western Australia, and the Australian Alps. It is probable that the more we look, the more evidence we will find. In 2016, researchers at Rosemary Island, in the Dampier Archipelago of Western Australia, established the age of one stone building at 9,000 years. It was built at the end of the last Ice Age, and is said to reflect how the people were adjusting to sea-level rises and the necessary movement of coastal clans inland.[45]

Foundations and walls are still visible, despite the pilfering of their stone for European dwellings and dry walling, two hundred years of damage by cattle and sheep, and the sudden advent of uncontrolled fires. There is much research still to be done on the use of stone by pre-colonial Aboriginal groups.

Prior to photography, one of the few pictorial records of such housing was a drawing by an Aboriginal artist of a group of stone buildings in the Australian Alps. This drawing is significant, because when a man from that region visited a large gathering of Aboriginal people in 1839, at what are now the Botanic Gardens in Melbourne, he was said to be the man who dreamed corroborree for the whole of south-eastern Australia. His home was a stone house.

His name was Kuller Kullup, and he refused to speak or even look at the Aboriginal Protector, William Thomas. Kuller Kullup was probably of the Jaitmathang or Ngarigo people, but all south-eastern Australian Aboriginal people knew him as a great philosopher.

Early travellers in the Alps remark on the small villages of stone houses and the large populations. Recent bushfires have uncovered the foundations of such settlements in various parts of Victoria, while numbers of other villages with stone walls and roofs of thatch have been recorded across Australia.[46]

Aboriginal Protector Herbert Basedow left detailed records of houses in the north-east of South Australia in 1925. He described houses roofed with flat slabs of stone, probably a local slate laid across timber beams.

Robinson's drawing of Caramut, a village in south western Victoria, c.1840

(Paul Memmott)

Dennis Foley, of the Gai-marigal people north of Port Jackson, remembers being shown the remains of large stone dwellings by his uncle. The walls were built from stone and clay, and the floors were covered in soft paperbark and ferns. The buildings were approximately six metres by four, and one-and-a-half metres high. Foley explains that many of these houses were burnt by soldiers after the smallpox plagues.

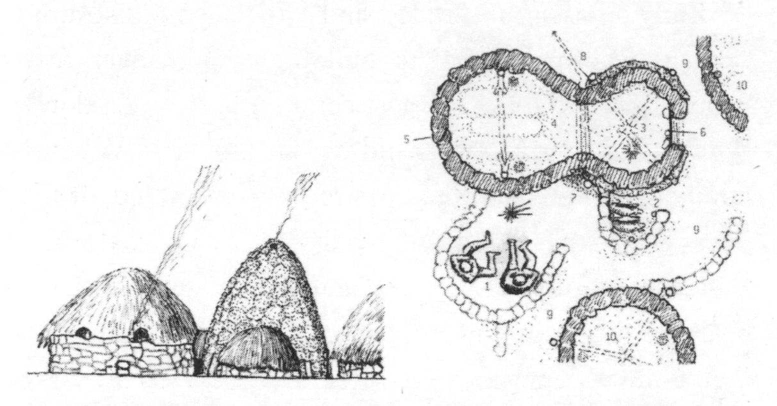

A reconstruction of a Gundidjmara village, western Victoria
(Paul Memmott)

The stone houses of Lake Condah and Tyrendarra in Western Victoria are among the best known as a result of the research undertaken by Heather Builth and members of the Gundidjmara clans. The Winda-Mara Aboriginal Cooperative conducts tours of the villages at Tyrendarra, and the Gundidjmara run cultural tours at Tower Hill. Apart from the remains of the houses themselves, the evidence is supported by early colonists such as James Dawson and Peter Manifold, who described the buildings in great detail.[47]

In some instances at Lake Condah, buildings shared walls for strength; some had interior doors. The roofs were often thatched by dense layers of grass or leaves, but many were protected by deep sods of earth, with the grass turned inwards. The central chimney was closed by a sod during periods of heavy rain.

Early reports from settlers and colonial administrators such as Robinson refer to buildings where over fifty people gathered, but the most common size was a dome three to five metres across and two metres high. When a family had more children, extra rooms were added, or the larger structures underwent subdivision by internal walls. Basalt is common in the area, and was used to construct both the fish traps and the houses. As mentioned earlier, the building materials were often commandeered by the squatters for constructing the network of basalt farm fences seen in the region today. Ironically, those fences are now subject to heritage protection.

The doorways of the houses could be closed by a bark-and-timber door, and when vacated, sticks attached above the door indicated to other clan members in which direction the family had gone. Robinson remarked on this message system. He drew a variety of the house styles during his travels as Chief Aboriginal Protector. Most appear to have had low, stone walls and a thatched or turfed dome. Some had an auxiliary curving wall for protection from the winds, giving the whole structure a floor plan roughly in the shape of the number six. The walled section seems to have been designed as a kind of courtyard.

Robinson's drawings are simple but evocative. He reported that the walls and roofs of the beehive, or kraal, type were so substantial that they were strong enough 'for a man on horseback to ride over'.[48] One wonders how that observation was proven and what the owners of the house might have thought about the experiment.

The great village and aquaculture complex at Lake Condah has been nominated for World Heritage Protection, and is a credit to the community's vision and persistence. It is now acknowledged as one of the world's significant sites of human development.

Readers interested in early Aboriginal housing would find that Paul Memmott's book provides great detail and illustration of their styles, and an indication of their ubiquity. For instance, he reports researchers finding hundreds of stone buildings on High Cliffy Island off the Western Australian coast. The island is only one kilometre long, but the high-density settlement was supported by intensive fishing and artefact manufacture from stone in a quarry. It might have been a seasonal camp, but the possibility exists that it was occupied permanently.

Other stone constructions

Builth noted other constructions, including two six-metre-long parallel walls built about a metre apart. The use of some of these structures is open to conjecture, but the complexity of these villages indicates the need for further enquiry. The researchers John Morieson and Lynne Russell reflect on the stone arrangements found throughout the country, and the determination with which colonial Australia tried to prove that they couldn't have been built by Indigenous people.

Morieson's work on the massive arrangements in various parts of Victoria led him to believe they were meant to predict the solstice. His mathematical calculations show precise correlations with celestial phenomena. Attracting attention and funding to such research is always difficult, but, as the evidence mounts, a new area of enquiry must surely develop.

Perversely, some early colonists exaggerated the size of some features in order to distance the structure from the capability of Aboriginal people and to suggest their origin to be the work of isolated Europeans in the distant past. An engraving of some vertical stones at Mount Elephant, Victoria, published in *Australian Illustrated News* in 1877, has certainly been exaggerated to assume Stonehenge proportions. Faced with the evidence of permanent occupation, some were even tempted to infer that the work was the result of aliens.

Denigration of Aboriginal building by Europeans took several forms. The explorer Ernest Giles was contemptuous of Aboriginal achievement. That prejudice squeezed his racism like toothpaste from a tube, so it is no surprise to read that when he discovered huge piles of stones with large central slabs, he assumed, with no evidence, that they were for cannibal sacrifice. If they were not old house sites, they were probably graves of significant figures, but it was too much to expect that Giles would describe them in anything but derogatory terms.

Photographs of other explorers and squatters with

their foot resting on a grave or house resemble the African big-game hunter claiming a rhinoceros as a trophy. The stance says everything about the opinion of the pipe-smoking trophy hunter as he assumes control of the country with so little conscience that it barely occurs to him to wonder at the nature of the civilisation that erected such monuments.

The underestimation of Indigenous achievement was a deliberate tactic of British colonialism. Large structures of North American First Nations people were similarly ignored, or credited to earlier Europeans; and in South Africa, Cecil Rhodes made it illegal for anyone to mention the huge Shona structures found in what was once Rhodesia and is now Zimbabwe. It was obvious that they were built by the Shona, but in order to legitimise his usurpation it was important that their achievements were denied.

The anthropologist Harry Allen commented at the 2013 History Symposium at the Australian National University that anthropologists have been singularly unsuccessful in combating the reductionist ideas of Australian people when thinking about Aboriginal Australia. He believes that many still see the continent with the same mind and eyes as the navigator William Dampier.

As discussed earlier, exaggerating the size of Aboriginal structures was used as an alternative colonial tactic. McNiven and Russell put it like this:

> [T]he Mt Elephant megalith sites helped European colonists legitimize their right to inherit the Australian continent. European colonization literally became a process of the (re)possession of a lost domain of their heritage … (it) was a strategy employed by European colonists in Africa, America and Australia to dissociate Indigenous populations from their pasts to establish a white connection as a prelude to appropriation under the guise of re-possession.[49]

In Australia, however, many of the Indigenous structures were built on a massive scale. Some stone arrangements needed no exaggeration, as some covered many hectares. These structures have been found all around the country, as have stone houses similar to the Lake Condah buildings.[50] Substantial dwellings were built from flat slabs of sandstone in the Kimberley region of Western Australia, along with a variety of shade houses to provide protection from the heat during the day; many were associated with stone fish-traps or stone designs of a purely religious nature.

All large buildings and villages had cooking ovens and food-preparation facilities. Some of these structures were so large and had been used for so long that they became raised above the surrounding land by several metres as a result of the build-up of ashes swept from the fire each morning. Robinson saw large ovens close to Melbourne even as late as 1841. Some were three metres wide. The ovens were used to cook food in identical fashion to the Maori hangi and the Papuan stone ovens.

Stone was also used to construct hides from which to hunt animals and birds, and as protection for sacred items. Ovens and grain stores were built by combining stones with clay mortar. Well covers were made from large slabs of stone ground down to fit neatly over the well in order to prevent animals and litter polluting the water. A huge disc of stone existed on one such well in the Victorian You Yangs right up to the last decade, but it was removed and rolled down the hill by vandals.

In 2009, I was shown a small but important well near Marree, South Australia, where small slabs of stone were interleaved to protect the well from animals and to reduce water loss through evaporation. The well was opened and closed with reverent ceremony.

Memmott describes buildings made of many materials, but most relied on the strength of the dome. In the south-east of the continent, whale bone was often used, as its curve produced great strength when radiating lengths were bound in the centre and thatched.

Memmott quotes the great Danish seafarer Jorgen Jorgensen's observations of such buildings on the north-west coast of Tasmania. Jorgensen was an intriguing character in his own right, later returning to Iceland, where he became monarch. He was impressed by the beauty and neatness of the buildings he observed in Australia. One of the 'beehive' shaped domed huts he noted was over seven metres in diameter.[51]

Marree well

(Lyn Harwood)

In 1974, Bill Mollison reported that 'the hut bases, cut into slopes, are visible today on the coast and give ample evidence of their size and structure'.[52] Some could accommodate fifteen people, and employed timbers bent under steam. Similar structures comprising whale-bone beams were reported as far away as the Great Australian Bight, where the interstices and roof were made of boughs and dried grass.

In the interior of the continent, whole spinifex plants were employed, both in windbreaks and fully enclosed dome structures. The clumps of spinifex were laid so that the clod and root ball of each plant met at the crown of the house, and the leaves extended down the sides. Sometimes the spinifex was coated with a clay render.

These waterproof claddings were observed by many, including Thomson, who remarked on the style after his contact with the Pintupi people.[53]

Jimmy Pike, a Walmajarri artist, described the process of cladding similar buildings in the Western Desert with grass and mud.[54] Architect Peter Hamilton adds an interesting anecdote about the Western Desert designs that may well apply to other regions. Hamilton speculated that the low-roof style tended to deter flies because of the darkness within. Many observers commented on this principle from other areas. Flies are a general nuisance in Australia, but many of the objectionable blowfly species were introduced into the country with sheep. Even so, the small black bush fly still knows how to make its presence felt.

Sacred design

Many design features had significant religious weight, and the lore of building is associated with creation stories. The names of building components often had dual meanings indicating these spiritual and physical functions.

Eastern Arrente people say they were taught the rules of building by a man of great wisdom. One instruction was the laying of foliage so that leaf fingers overlapped leaf fingers and thus shed the rain. This story was re-told by Walter Smith (Purula), who provided the commentary on sowing and irrigating grain.[55]

As a consequence of the mythological association with

building styles, certain symbols, such as ridge poles with an attached skirt of thatch, and the forked pole to support the ridge, are central features of some contemporary ceremonies and art. As Memmott indicates, the detail of these symbols is of a sacred nature, but you can imagine the symbolism that forks and poles might encourage.

Markings near Sutton Forest, New South Wales c.1899

Villages and cemeteries were often marked with carved and painted tree trunks and timbers.

(Michael Young)

Burial within cemeteries is another of the indicators of sedentism recognised by archaeologists, and abundant examples are provided in explorers' journals.

A variety of structures were erected over graves in many different parts of Australia, and where cultural practices have survived this element can still be observed.

Many of the photographs we have of graves and graveyards indicate that the site was selected for its beauty, and many show arbours than we can assume had been planted to enhance the aesthetic.

Landscaping and garden design around cemeteries and ceremonial grounds were observed by a number of early settlers and explorers, and evidence of some of this work remains today in places as remote from each other as Maningrida and Cape Howe, in eastern Victoria.[56]

In 2010, Neville Oddie, a fourth-generation central Victorian farmer, showed me a series of ring trees on his property near Ballarat. A ceremonial ground stood at the edge of an area of native grasses. Family records show that this area has never been ploughed, and he has continued the family tradition.

Oddie was keen to preserve the original, pre-colonial grassland and the ceremonial grove that overlooked the plain. Trees within the grove had been altered by lacing one limb over another while the trees were still saplings, so that as they grew the limbs fused and left oval-shaped windows or rings. Some have speculated that there is a sexual element to the design. I have seen other intriguing examples of this kind in other parts of Victoria, and have wondered about the intervention of Aboriginal people in the design of these arbours.

Jane Pye from Walgett lives on a property held by her family for several generations. The family took pride in preserving the many altered trees on the land, and have continued to include local Aboriginal communities in that preservation. The photos below are a minute sample of the hundreds of altered trees on the property.

There is a design feature in some of the canoe trees where there is a distinctive bulb at either end. Recently, I saw a very similar feature on a tree at Bermagui on the New South Wales south coast.

Altered trees, Walgett

(Jane Pye)

A grove of mahogany gums at Penders in New South Wales suggests a degree of deliberate horticultural shaping, and a bora ground near Lake Barracoota on the Victorian and New South Wales border indicates that the area was a planned ceremonial grove. It is overgrown with scrub now, but if you use your imagination, the aesthetic is obvious. I included both sites in my novel *Bloke*, because the harmony of each was so striking.

Thomas Mitchell's journals are important records of ceremonial burials, not just for the field notes, but because of the drawings accompanying them. His drawing of the cemetery at Milmeridien, near the Darling River, is a thing of great beauty and evokes serene repose:

> The burying-ground was a fairy-like spot, in the midst of a scrub of drooping acacias. It was extensive, and laid out in walks, which were narrow and smooth, as if intended only for 'sprites'; and they meandered in gracefully curved lines, amongst the heaps of reddish earth, which contrasted finely with the acacias and dark casuarinae around. Others gilt with moss shot far into the recesses of the bush, where slight traces of still more ancient graves, proved the antiquity of these simple but touching records of humanity. With all our art, we could do no more for the dead, than these poor savages had done.[57]

You cannot look at Mitchell's drawing and maintain a belief in the brutish description of Aboriginals that Australian history insists we accept.

Mitchell's drawing of the cemetery at Milmeridien, near the Darling River
(State Library of South Australia)

Bough shelters or low awnings are still seen over graves in Arnhem Land's Maningrida in a ritual associated with the stages of mourning following interment. Interestingly, Memmott's book includes a spatial plan of the Maningrida township from 1970; but, even though it has grown to a town of 3,000 people, that same town plan is discernible today, as it is based on the spatial separation of the component language and cultural groups. It is intriguing to note that in 2010 Maningrida had nine football, three soccer, and several other sporting teams playing vibrant competition. It is not uncommon to see four full-scale games played within seven days, with training on the other nights, and bands practising modern and traditional music afterwards. A very lively and interesting community.

The modern township of Maningrida is a fascinating example of external living spaces associated with houses. All Aboriginal communities, whether they built stone-walled thatched domes or more open shelters, had reserved spaces outside the building for sleeping and resting during fine weather.

Social interaction between groups is crucial in sustaining communal life, and outdoor living spaces encouraged observance of and interaction with the wider community. It is common to see people sitting in front of their houses watching the life of their community pass by. Even in urban areas of Victoria, this practice is maintained, making it very hard to pass by an Aboriginal dwelling without being noticed.

The studies of living spaces conducted by people like Memmott, Thomson, and Hamilton indicate the importance of the articulation between inside and outside spaces, a point not always considered in the design of contemporary Aboriginal housing programs.

This has been just a small review of traditional Aboriginal construction techniques; while many of the examples are simple, many others were complex and permanent. The importance of examining this material is to dissuade a common Australian perception that Aboriginal and Torres Strait Island people built nothing more complex than a piece of bark leaning on a stick.

Mitchell's drawing of tombs

(State Library of South Australia)

Blandowski's interpretation of a burial place described by Sturt

(Haddon Library of Archaeology and Anthropology)

The reason for the knowledge of complex building techniques being dropped from the Australian conscience is not as important as the urgency for its re-incorporation into our contemporary consideration of the traditional economy. While we continue to think of Aboriginal people as having no construction skills, it is easier to dismiss Aboriginal attachment to land. Moreover, the insistence on using the hunter-gatherer label is prejudicial to the rights of Aboriginal people to land.

4

Storage and Preservation

Pottery is one of the tests applied by Western archaeologists to the developmental level reached by civilisations. Australian Aboriginals would, at first glance, appear to have failed this test. The superb glazed and kiln-fired pottery of China, Greece, or Rome has not been found here; however, clay vessels were made. While most were relatively crude sun-dried bowls, some were baked beside the fire; others, particularly small clay figurines, were fired on charcoal beds, and some were glazed with mineral washes.

The tests applied in this way simply check how similar a group is to European and Asian civilisations, and may not reflect their success in other areas such as social cohesion, resistance to warfare, or sustainable use of resources.

This chapter looks at elements of Australian pottery and food preservation, because the perceived lack of them has been used as an indicator of social backwardness. This attitude prejudices opinion about the level of development of Aboriginal and Torres Strait Islander people. To point out that Indigenous Australia did indeed use baked-clay vessels and preserved food is not an attempt to claim distinction for the First Australians, but simply to point out that if this were the only test for development, it cannot be seen as completely absent from this country.

If the test of sophistication were whether or not all were fed regardless of rank, or whether all contributed to the spiritual and cultural health of the civilisation, Aboriginal Australia might have a much higher rank than some of the nations considered the hallmark of human evolution.

Storage containers, constructed from an enormous variety of materials, were found across the country. Water vessels made from animal skins or intestines were common possessions. The southern coastal populations crafted smaller water carriers and pouches, and these may have been more useful when carried for any great distance. Water carriers made from bull kelp in Tasmania are items of enormous beauty. Light shines through them with an amber glow. That they are unknown to most Australians is not an indication of their utility or grace.

According to the earliest records, the use of clay to render houses or make storage vessels was witnessed in most parts of Australia, although the crude drying and firing methods may have resulted in the remaining fragments being overlooked by later surveys. Stores of food were seen across the continent, too. Although most disappeared quickly, some ossified caches have been found in stone chambers, preserved by the tight-fitting stone plugs.

Gerritsen suggests that the storage of food surpluses is one of the indicators of agricultural nations, and defines three types of food storage used in Australia: 'caching, stockpiling and … direct storage'.[1] Caching he defines as small stores preserved in some manner and left in a protected location. As an example of caching, he refers to the Kukatja and Pintupi of the Great Sandy Desert, who harvested acacia and eucalypt seed, and covered them with spinifex for consumption later in the year when all other foods had been exhausted.

Stockpiling was most commonly used prior to large ceremonies where hundreds of people were catered for over an extended period. Examples of such stores were seen by many explorers, and were utilised by them to sustain their progress.

Direct storage, by Gerritsen's definition, was evidenced by the chambers made of clay and straw for storing 'seeds and fruits, nuts, gum, tubers of various sorts, eggs, meat, fish, fish oil and even mussels'.[2] Large grain stores of more than 50 kilograms were also found in perfect condition sewn up in animal skins. Hollow trees and rock wells were also used.

Skin bags were frequently used to store grain and other produce. The explorer Charles Coxen found 45 kilograms of grain near the Castlereagh River, and Howitt found 50 kilograms in the Poeppel Corner. On the Barkly Tableland, Ashwin discovered a complex store in a settlement of fifty huts surrounded by a fence 180 metres in diameter, where 'a large mia-mia about 7 feet (2.1 metres) high in the middle and about 16 foot (4.8 metres) diameter stored 17 large wooden dishes four or five feet (1.2–1 m) long filled with grass seed as large as rice'.[3] The store represented a ton of seed. The finds of tons of grain by Sturt, Giles, and others have already been mentioned (see chapter one).

Howitt, when searching for Burke and Wills, described one of the grain stores:

> Near Lake Lipson, one of my party found about two bushels contained in a grass case daubed with mud. It looked like a small clay coffin and was concealed … the munyoura [munyeroo, nardoo] bower tastes like linseed-meal, and is by no means unpleasant when baked in ashes and eaten hot.[4]

All manner of foods were preserved and stored, but such preservations became more difficult when the clans were forced into constant movement by the advance of pastoralists.

The variety of preservation techniques was remarked

upon by the earliest European witnesses. Gums were preserved in flat cakes, and milled flour was patted into large round balls and then dried for preservation. Stores of fish meal and fish flour were recorded, but many other commodities had their individual preparations prior to storage, including caterpillars, witchetty grubs, grasshoppers, meat, and liver. Such stores were often coated in the ashes of particular woods, and later mixed with seed flour before cooking.[5]

The science of food preservation and treatment allowed Aboriginal people to render otherwise toxic foods edible. The pandanus and cycad nuts went through stringent sluicing and immersion treatments to remove the poisonously high alkaloid levels.

At certain times of the year, some yams were bitter with alkaloids, but there were solutions. The following recipe was one of many used to solve the problem: 'Partly cook the tuber. Slice into discs or wedges. Coat slices in wet ashes of river gum until covered in paste. Cook overnight in a good oven.'[6]

The nardoo (*Marsilea* sp.), which so confounded Burke and Wills, is a plant whose thiaminase content is so high that it has to be carefully sluiced before further preparations, which included pounding, winnowing, and baking. Thiaminasc blocks the absorption of Vitamin B, and this may explain the explorers' failure to thrive on the food. Perhaps an explanation of the required techniques may have been forthcoming to Burke's doomed party if he had refrained from firing his pistol at the people who were trying to keep him alive.

The greens from the top of the nardoo plant were eaten as a steamed vegetable. Little wonder that the plant features prominently in Aboriginal culture.

Another plant that is toxic if left untreated is the burrawang (*Zamia spiralis*). It was roasted and pounded, and then the mash was left in water for two to three weeks to remove toxins. The early horticulturalist Ansthelme Thozet said the matchbox bean required similar complicated preparation.

Some plants were so prolific that they provided food for large gatherings of people, not only during the harvest, but later when stored quantities could be eaten. Bunya nuts, which come from a conifer in the genus *Araucaria*, are an example of a plant that fruited so heavily that large stores were set aside.

The great gatherings of people in the Australian Alps were made possible by the summer arrival of massive numbers of Bogong moths. The Alps were a crucial hub of cultural knowledge, which was disseminated to all the surrounding clans, and this seasonal feasting must have been an intriguing political and social event. Maneroo, Bidwell, Ngarigo, Yuin, Thawa, Diringanj, Walbanga, and the Ngunuwal from Canberra were just some of the clans to attend these harvests.

The moths were collected in vast numbers from crevices in the rock, and either swept onto nets made of kurrajong fibre or whisked onto kangaroo skins. The moths were cooked in hot ashes for a short time until the wings and legs were singed away. The moth carcases were placed on a bark platter until cool, then collected and sifted in a net until the heads fell off, after which the body was eaten or ground into a paste and made into doughy cakes, which were smoked to preserve them.

William Jardine, a Cape York pastoralist, observed that the moths were placed 'in an oven made of the burning sand, covering them up at once; in a few minutes they were cooked and when taken out looked like a beautiful white kernel, with a flavour of marrow'.[7]

Crows assembled to take part in the feast, and they too became fat and so intent on their hunt that the tribes knocked them on the head and ate them, a great delicacy, as the meat was plump and aromatic after the birds' diet of moth fat.

During the moth harvests, the moth was treated carefully, and if the body of the moth was scorched during cooking, a great storm was said to arise and blow moths off course and out to sea, with the bulk of the harvest lost. Such storms had been witnessed by white settlers, and these events would have caused great hardship for the moth hunters.[8]

Cooking, storage, and food-handling methods were governed by strict protocols and religious observation. In the north of the continent, harvests of yam had to be undertaken according to strict rules. The yam plant was not to be excessively damaged, and the tuber not bruised, or certain penalties would apply to the harvester. Yam daisy harvests in the south were attended by similar protocols, all ensuring the protection of the plant.

The richness of this diet is attested to by settlers who witnessed Aboriginals returning from the moth harvest in great health, their bodies glistening with moth fat.[9] The ability of the crop to sustain such huge numbers of people over the harvest season is explained by the fact that 50 to 60 per cent of the moth's weight is fat.

Storage vessels of various kinds were required for the storage of these foods, but another example of the use of moulded materials was discovered in the last decade. A site was discovered in Central Australia that contained dozens of spherical vessels made from gypsum or similar material.[10] These artefacts had been observed by Basedow and others in many parts of Australia, but they are notoriously fragile, and frequently did not survive the wave of sheep and cattle moving across the land or the burning of Aboriginal villages by squatters. Those vessels were mourning caps, or widow's caps, and were worn by Aboriginal women after the deaths of their husbands. They were made by taking a cast of woven cord, and covered in kopi, a white gypsum clay.

Widow's caps have been found in the Simpson Desert near Birdsville by two Wangakurru Elders who were retracing storylines told to them by their Elders. This site contains forty caps as well as large grinding stones, and indicates that the grave belonged to a very senior Aboriginal person. These caps are of such a quality that they have survived 200 years in a harsh environment. They are frequently depicted in pre-contact art and post-contact photography. There is a possibility that the caps have been made in more recent times by people maintaining cultural traditions into the post-colonial period, but nevertheless they survive as crafted objects associated with the early stages of pottery.

Gypsum caps — another example of the use of moulded materials
(Haddon Library of Archaeology and Anthropology)

Maree Clarke of the Koorie Heritage Trust in Melbourne featured these objects and the ceremonial performances associated with grief in her 2012 exhibition.

Her images are not only riveting, but a timely reminder that Australians have only scratched the surface of Aboriginal knowledge and manufacture.

Gerritsen quotes Officer, who observed that some ceremonial objects were made out of gypsum, 'which is first burnt and then mixed to form a cement with sand and water, moulded to the required shape, and afterwards evidently finished by scraping'.[11]

Clay and stone cooking ovens use all the principles of pottery, and some burial repositories employ the same skills. Similarly, water wells were often lined with baked clay, fired while the well was dry.

Rectangular troughs with sealed lids have been found full of grain, and many caches of grain and other foods were stored in grass packages, which were then smeared with clay to form an impermeable vessel.[12]

Clay coating of house roofs deflected rain and maintained more moderate temperatures inside the dwelling. These veneers were sun dried, and of such strength and durability that George Augustus Robinson claimed you could ride a horse over them.

It would seem that Aboriginal and Torres Strait Islander civilisation was on a trajectory towards greater and more sophisticated use of pottery, but many of the societies claimed by anthropologists to have left the era of hunter-gathering and joined the march towards agriculture never used any form of pottery.[13]

We have to be careful that we are not deciding on

markers of civilisation simply because that is the historical path followed by Western civilisations. As Gavin Menzies, author of *1421* has pointed out, if you proceed on the assumption that only Western European nations had reached the stage of civilisation, you have to behave as if the Chinese were not the first to invent gunpowder, pottery, and celestial navigation techniques.

China was probably the most advanced nation on earth until the eighteenth century, but arrived there without following all the steps that Westerners consider the true path to civilisation. Racial bias can cloud observation and reasoning.

The same intellectual prejudice has been applied to the state of Aboriginal civilisation in Australia, and in doing so has dismissed large bodies of first-contact evidence as being mere aberrations. With that myth firmly planted in the minds of students, it was only a matter of time before the image of Aboriginals as primitives haplessly wandering across the face of the earth allowed Australians to feel 'sorry' for Aboriginals and dismiss them from the national consciousness.

Some modern archaeologists see archaeology as simply another colonial strategy designed to justify occupation of 'savage' lands by western countries. Hutchings and La Salle say that, '[A]rchaeology has always been tied to the endeavour of imperialism and capitalist expansion ... Such "proof" of the natural, cultural and racial inferiority of Indigenous peoples served the interests of those who would justify their subjugation, assimilation and enslavement for profit.'[14]

They quote McNiven and Russell's assertion:

> [A]rchaeologists and prehistorians constructed the archaeological record to scientifically vindicate the colonialist notions of savagery and staged progressivism to leave little doubt that Indigenous peoples, particularly 'hunter-gatherers', represented primordial man ... [and that archaeology] has little to do with the rigors of science and all to do with a colonial ideology and a ... public that wishes to find scientific support to legitimize colonial dispossession of Aboriginal lands and to delegitimize contemporary Aboriginal claims to Native Title rights.

Many Australians are galled when Aboriginal and Torres Strait Islanders urge a revision of Australian history, but Aboriginal people are forced into a defence of their history, culture, and economy every day of their lives.

The following may serve as an example of how assumptions of Aboriginal inferiority affect contemporary Australian reconciliation.

In 2009, I bought my wife a holiday so she could fulfil two lifelong ambitions: to see North Queensland Aboriginal art, and turtles coming ashore to lay their eggs.

An Australian organisation that produces a very valuable quarterly magazine showcasing the country's geography offered a package that seemed ideal. It wasn't cheap, but I was thrilled to be able to give my wife a holiday to remember. We were promised experts in the fields of art, science, and natural history. On the first evening, I was listening to one of the experts re-tell his adventures on a 4WD trek through the Kimberley. I let the fascination with 4WD bravado go through to the keeper.

As the recollection rolled on, I was stunned into silence. The guru of Aboriginal art proceeded to boast of how he had duped the local Land Council and gained access to restricted parts of their land.

Our great Australian geographers would never think of breaking into the Pine Gap military installation, or wandering unescorted through the Pilbara iron ore mines; but, on receiving a polite refusal to their request to enter the culturally restricted zone, they went to the local police.

The police were galvanised into action, relishing the opportunity to thwart the authority of uppity blacks. *If we perceive a crime has been committed*, they told our adventurers, *we can go where we like. We perceive a crime*, they chortled.

So the police escorted the 4WD heroes into the initiation site. Once there, they threw beer cans into the sacred water, and took it in turns to shoot the cans with police-issue Glock pistols.

The 'explorer' gloated over his win against the Land Council, which for many Australians may seem mere cheekiness. But when next the local Aboriginal Elders brought their young men to the initiation site, they found it full of bullet-riddled beer cans.

The most disturbing thing about the event was that it undermined the authority of the Elders. They were trying to impress on their young men the importance of maintaining culture and a responsible, alcohol-free way of life. The young men would have seen immediately that Australia had no regard for the authority of the Elders.

I was stupefied by the contempt displayed in this story, but didn't tell my wife because we were three days from the turtles. Surely I could bite my lip for that long.

Two nights later, we were sitting around the communal fire when the art guru derided Kimberley art and culture, claiming that the Bradshaw paintings, a fine group of rock paintings recorded by the pastoralist Joseph Bradshaw in 1891, were the work of Asian people because they were far too beautiful to have been painted by Aboriginals.

Mike Morwood, the archaeologist credited with the discovery of the 'Hobbit' skeletons on Flores in Indonesia, is conducting intensive research on the paintings. He believes the paintings to be 40,000 to 45,000 years old, and may have been the work of an earlier migration to Australia. However, significantly, he points to similar art in Arnhem Land, where a fallen panel of art offers the opportunity to test the sand beneath it with the thermoluminescence technique to reveal the age when the slab fell.

The new analysis will be fascinating, but the existence of art at other Australian sites, where the depictions show similar ceremonial dress, seems to refute the idea that the early Kimberley painters were a different race.

Professor Peter Veth of the University of Western Australia acknowledges the debate, but says:

> [T]he suggestion of culture demise is not comprehensive and is not supported … however we do have evidence of climate change and people signalling very differently in their art [but] … a change of art style does not equate with a new people … graphic switches occur in the art of Aboriginal Australia in many places.[15]

Indeed, graphic changes in art style occur in any art community all the time.

I'd heard all the 'superior civilisation' theory before, but even in the previous month's issue of the tour group's own magazine there had been a lengthy article about the provenance of the Bradshaws, including a refutation of the misunderstanding by white 'experts' when told by local Aboriginal people that 'We did not make these paintings'.[16] No, their ancestors did. This is too difficult a concept for many art experts, and they leap on the idea that the art is the work of a more sophisticated people. What finer way to denigrate Aboriginal culture?

I tried to point out to the 4WD cowboys that the

University of Western Australia and their own magazine had dismissed such nonsense. But the 'experts' were not to be denied, and shouted us down. We left, and toured on our own.

We did see the turtles hatching, and we did spend two wonderful days touring the Laura art sites. We also spent time with family at Lockhart River, but the experience with the white experts burnt. Humiliation always does.

5

Fire

The use of fire has always had a central place in the Australian imagination, but this was further embedded after the horrific fires in Victoria in February 2009. The loss of 173 lives, with 414 people injured, 2,029 houses destroyed, and $1.5 billion in lost property turned the country's attention to the role that fire plays in the national psyche. We are terrified of fire.

But it wasn't always like this.

Palynologists such as Singh and Kershaw have evidence which they suggest supports the fact that Aboriginal Australians began using fire as a tool over 120,000 years ago — even though most archaeologists believe human occupation of the continent occurred no earlier than 60,000 years ago.

Tim Flannery, in his book *The Future Eaters*,

suggested that the Aboriginal use of fire contributed to the extinction of Australian megafauna 40,000 years ago. Jim Kohen, from the School of Biological Sciences at Macquarie University, points to the fact that megafauna survived in some areas into the Holocene, which began 12,000 years ago. He suggests that the use of fire by Aborigines caused gradual change to vegetation zones, and that these changes are reflected in tool technology. As grasslands developed and the megafauna gradually disappeared, Aboriginal food production shifted from hunting big game to smaller game, and to an increased reliance on grains and tubers.

Kohen suggests that spear-point technology changed dramatically about 2,000 years ago in order to hunt smaller game, including fish and possums. Increasing reliance on grains and tubers initiated tool innovations, including the juan knives, which Gregory saw being used to harvest grain. The increase in the production of adzes for sharpening wooden instruments such as digging sticks indicates a shift towards the more intensive cultivation of the yam daisy.

As mentioned earlier, some researchers believe that the so-called intensification period began much earlier in Australia. Innovative thinking and investigation has begun into the land use of Aboriginal people by researchers such as Rupert Gerritsen, Bill Gammage, Beth Gott, Jeanette Hope, Harry Allen, John Blay, Tim Allen, and others, and their work will challenge almost everything scientists have so far assumed about Australian pre-colonial history.

Early anthropologists and historians assumed that the continuous firing of the bush was a simple method of providing green pick to attract game. However, recent investigations, such as those above, and a review of explorer observations, show a much more complex operation.

Bushfire disasters, such as the Black Saturday fires in Victoria, generally centre on the wet sclerophyll mountain ash forests, but tree-core analysis indicates that wild fires in these forests were largely unknown before the arrival of Europeans. The means of management of these forests by Aborigines is not clear, but first-hand accounts of settlers and explorers indicate that a mosaic pattern of low-level burns was the method employed. The better soils were used for production, while the inferior soils were left for forest.

The areas around these forests seem to have been in a continuous state of management. Almost all early European visitors to Australia remarked on the frequency of small-scale burning. Gott quotes Thomas Mitchell on the subject:

> [W]here a man might gallop whole miles without impediment and see whole miles before him … the omission of the annual periodic burn by natives of the grass and young saplings has already produced in the open forest lands nearest to Sydney thick forests

> of young trees ... Kangaroos are no longer to be seen there, the grass is choked by underwood; neither are there natives to burn the grass, nor is fire longer desirable among the fences of the settlers.[1]

Mitchell hit upon the impediment that inhibits control burns in the Australian landscape today: farm fences. Fencing is expensive, as are the outbuildings for stock, hay, and equipment, not to mention private housing, and power and irrigation lines.

Governor Phillip observed in 1789 that the forest trees grew at least twenty to forty feet from each other and that there was very little undergrowth.[2] As early as 1827, Peter Cunningham described the countryside of Parramatta and Liverpool as lightly timbered and so clear of undergrowth that you could ride a gig in any direction without hindrance. The first European settlers were reminded of the manicured parks of England.[3]

Within years of the Aborigines being prevented from operating their traditional fire regimes, the countryside was overwhelmed by understorey species. Old settler families of north-east Gippsland have told me that when their forefathers were shown the country by Aboriginal people in the 1840s, all the plains were clean and well grassed, including the narrow river valleys. Looking at those valleys today, it is almost inconceivable, and modern, long-term farmers of the district are incredulous when they are told what their farms used to look like.

Norman Wakefield, an Australian palaeontologist and botanist, recorded the memories of one of the old-timers, J.C. Rogers:

> [I]t had been the accepted thing to burn the bush, to provide a new growth of shorter sweet feed for cattle ... The practice was to burn the country as often as possible ... in the hottest and driest weather in January and February, so that the fire would be as hot as possible, and thus make a clean burn [but] the long followed practice ... resulted in a great increase of scrub in all the timbered areas ... The fires forced the trees to seed and coppice, and in time an almost impenetrable forest arose.[4]

Changing the timing and intensity of fires radically changed the nature of the country, so that what had been productive agricultural land became scrub within a decade.

Kohen summarises this situation:

> While Aboriginal people used fire as a tool for increasing the productivity of their environment, Europeans saw fire as a threat. Without regular low intensity burning, leaf litter accumulates, and crown fires can result, destroying everything in their path. European settlers feared fire, for it could destroy their houses, their crops and it could destroy them. Yet the

> environment which was so attractive to them was created by fire.[5]

Rhys Jones is even more emphatic:

> What do we want to conserve, the environment as it was in 1788, or do we yearn for an environment without man, as it might have been 30,000 or more years ago? If the former, then we must do what the Aborigines did and burn at regular intervals under controlled conditions.[6]

Aboriginal people also had to protect housing, sacred sites, water courses, and the lands of neighbouring clans, but they were much more flexible in their planning. The crucial difference between the use of fire prior to the colonial period and since is the intensity of fire and available fuel loads.

The Aboriginal approach to fire worked on five principles. One, the majority of the agricultural lands were fired on a rotating mosaic, which controlled intensity, and allowed plants and animals to survive in refuges. Two, the time of the year when fires were lit depended on the type of country to be burnt and the condition of the bush at the time. Three, the prevailing weather was crucial to the timing of the burn. Four, neighbouring clans were advised of all fire activity. Five, the growing season of particular plants was avoided at all costs.

Examples of this studied approach can be seen in the advice given by Aborigines to Europeans when it became obvious that Europeans were using fire too infrequently and in the wrong conditions. The convict emancipee Robert Alexander was given very specific instructions by Jinoor Jack of the Bidwell-Maap people on how and when to burn the bush in the Genoa Valley in East Gippsland. In the months of February or March, he was advised to burn 'after the longest day when the sap begins to go down. In that period there are westerly winds in the morning that change to northeast in the afternoon, which provide natural back burn'.[7]

Alexander was told that this had to be repeated every five years. Many early settlers comment on how open the bush was and how travel was so much easier than when re-growth occurred after the fire regime was interrupted. It's interesting that in the Genoa region, morning dews and shorter days begin around the time of year suggested by Jinoor Jack.

Each area had specific requirements. The yam districts, for instance, were burnt after the plants had shed seed, and the tubers were dormant. Edward Curr wrote that the interval between burns was about five years. Donald Thomson observed burns in Arnhem Land in 1949, and remarked on how they were strictly controlled by Elders. Other commentators, including Mary Gilmore, A.P. Elkin, and Donald Thomson, noticed that the burns were part of the spiritual communication with the land.

Sections of the bush were burnt periodically to promote lush growth, in what most commentators assume is an attempt to attract game. That is probably one of the peripheral functions, but more and more evidence indicates that fire was part of a planned program of cropping or, as some researchers refer to it, fire-stick farming.

The disappearance of the yam daisy is directly related to the post-colonial fire regime. The introduction of sheep and the exclusion of controlled fire withdrew several crucial elements of the plant's ecological requirements. The only yam plants to be found today are on railway verges and other lands fenced off from livestock, and where no superphosphate has been used.

A research group comprising Aboriginal and non-Aboriginal people of Far East Gippsland spent a long time searching for remnant yam daisy plants in Eastern Victoria without much success, until John Blay and members of the Eden Aboriginal Land Council found a patch on the Bundian Way, in the Australian Alps. Since then, we have found several more, and greater awareness will bring others to light.

Blay suggests that plants such as the Vanilla Lily (*Arthropodium milleflorum*) may have been of equal importance as a staple source of starch and protein in this area. Planting trials of yam daisy are into their fifth year, and the data collected will allow the group to make an assessment of yam and other tubers in the traditional East Gippsland diet. The aim is to try to analyse a significant part of the country's Aboriginal economy.

Gerritsen talks about the domestication of plants, the modifications when human intervention is persistent. The domestication of the yam creates a reliance on human activity following the harvest of its roots over thousands of years.[8] It would be interesting to know if this domestication is now in reverse as a result of the declining plant populations, reduced carbon because of the absence of fire, more compacted soils as a result of the withdrawal of Aboriginal cultivation, and the cessation of harvests.

Similarly, the status of wallaby and kangaroo grasslands is crucial to our understanding of the nature and the economy of the pre-colonial diet in this area. Surveys conducted at Marlo and Mallacoota in East Gippsland in January 2012 showed large areas of these plants flourishing in coastal heaths following a controlled burn the previous year.

One naturalist swept his arm across the hip-high heads of kangaroo grass, and was surprised at the handful of grain harvested so simply. We need to go back to the archaeological collections from this area to examine the tools to see if some were related to harvests of these grains.

Several of these grasslands were harvested in the summer of 2017, and converted into flour. The baking and restaurant industries, having tasted the breads made from these flours, are clamouring for produce. Not only that, but these plants give us another method of reducing greenhouse gases, as these grasses are perennial — their massive root systems are able to sequester carbon. Supplementary advantages are that the land does not need to be ploughed to sow down crops, and the use of tractors and diesel fuel is reduced, both emission-friendly innovations.

The grasslands were burnt, but lightly, so as not to penetrate the soil and deplete it of nutrient. The fires were controlled, and carbon emissions minimised.

Harvesting at Mallacoota Airport 2017

(Lyn Harwood)

Winnowing and separating, Gipsy Point 2017

(Helen Stagoll)

The importance of fire to Aboriginal agriculture was evidenced by the fact that after the fires of Ash Wednesday in 1983 there was 'a phenomenal flowering of tuberous perennials'.[9] These had adapted to the horticultural intercession of man and fire, and had become a crucial part of the plant ecology.

Daryl Tonkin, a long-term resident of the country near Drouin in West Gippsland, remembers the catastrophic fires of 1939, which he attributed to the increasing reluctance of the Europeans to burn, and the habit of leaving the heads of felled trees unburnt.

Even so, the regeneration after the fires was incredible: 'A fire is good for a forest, the seeds cannot germinate without hot ashes to cover them. In the old days, the blackfellas who lived in the bush looked after the animals and birds by burning the bush for them ... Plants that have been dormant for years will grow after fire.'[10]

As recently as 1983, autumn fires at Anglesea, in south-western Victoria, resulted in an incredible flowering of tuberous plants the next spring.[11] Grasslands are known to benefit from periodic burns, but little recognition has been paid to the role fire played in the promotion of the tuberous plants that were staples in the Australian Aboriginal and Torres Strait Islander diet.

Beth Gott believes the first Europeans in Australia witnessed a changed and managed land that was neither pristine nor wild. The burns would certainly have determined the nature of the country, and would have favoured important food plants. She quotes Bowman's remark that 'fire was a powerful tool that Aborigines used systematically and purposefully over the landscape … (there is) little doubt that Aboriginal burning was skilful and was central to the maintenance of the landscapes colonised by Europeans in the 19th century'.[12]

The use of fire was so controlled that belts of trees separating grasslands were maintained, and even small copses were allowed to remain in an open plain by the judicious use of back-burning to protect them. The Aborigines were using fire to produce associations between plains, forests, and copses. It was planned and managed to enhance returns for their economy.

Bill Gammage quotes John Lort Stokes and Edward Curr respectively: 'The dexterity with which they manage so proverbially a dangerous agent as fire is indeed astonishing.'[13] And, '[T]here was another instrument in the hands of these savages which must be credited with results which it would be difficult to over-estimate. I refer to the *fire-stick*; for the blackfellow was constantly setting fire to the grass and trees … he tilled his land and cultivated his pastures with fire.'[14]

Tim Flannery considers that Aboriginal people became 'ecological bankers in the Australian environment'.[15] He points to the work of Elinor Ostrom, the 2009 Nobel laureate in economics, who believes that, under certain conditions, chiefly the ability to exclude outsiders and a reliance on mutually agreed rules, humans can manage what she called 'the commons' in a mutual and sustainable fashion.

Flannery believes Aboriginal society fitted these conditions:

> As the term firestick farming suggests, the Aboriginal use of fire resembled agriculture in some ways: it yielded certain crops at certain times, suppressed weeds and was carefully controlled … Aboriginal people are fiercely protective of their clan lands, excluding outsiders or inviting them in as conditions warrant. There are also clear rules about who has the right to what resources and highly evolved mechanisms to resolve conflicts and enforce penalties. This has enabled Australia's Aboriginal people to act as keystone species of the continent's ecosystems for forty-five thousand years. As the Europeans displaced them, Australia's fragile environment collapsed into a far less productive and diverse state.[16]

Bill Gammage has made an exhaustive study of the Australian Aboriginal use of fire and noticed that, even in contemporary times, grassland production has been used as a lure to kangaroos and emus but, primarily, to keep stock away from deliberate plantings of grain and tuber crops. He speculates that the positioning of wells between kangaroo pastures and croplands provided all the needs of animals so they had no cause to enter areas dedicated to crops. A psychological fence.

Certainly, the creation of grasslands favourable to kangaroos, other smaller mammals, and emus allowed Aborigines to locate and utilise game. The discerning use of fire could produce sweet green feed in one area, and leave another covered in rank and dry feed. Fire was used to determine where animals would congregate.

Gammage maintains that fire was used:

> To shape the land ... It was a major totem, a friend. People knew when to use it and when not to. They knew if they released it according to universal law and local practice it would do what they wanted. If it did not then they, not it, had offended ... Like songlines, fire unified Australia. It locked the landscape into long-term widespread patterns, because neighbours obeyed the same law, and co-ordinated their burning or non-burning.[17]

He argues that Cape York people burnt patches of ground six to seven kilometres apart in recognition that a mob of disturbed kangaroos would only travel that distance. So if a mob left one district, the Aborigines would know exactly where to find them.

In *The Biggest Estate on Earth,* Gammage shows that the best lands were used for pasture and crops, and the inferior soils for forest. The land was broken into mosaics of cleared land and forest, not as part of some random fire regime, but as part of a plan to maximise food resources. He refers to these mosaics as templates, and explains their complexity: they helped protect individual crops, provide shelter for towns, and improve the aesthetics of living zones.

Colonial artists have been accused of romanticising the Englishness of the landscape, of falsely representing the land in their early paintings.[18] Surely, art critics and historians argued, the park-like country depicted was a result of the artist's nostalgia and his unfamiliarity with the country? Gammage visited the locations of some of those iconic colonial paintings, and found that they are generally true to their subject, even to the recording of particular rocks and trees still identifiable today by their unique arrangement.

The Port Phillip surveyor Robert Hoddle and others were at pains to emphasise the deliberation required to create this landscape aesthetic. The patterning by the use of scrub-eliminating forest created a series of attractive mosaics, or, as Gammage refers to them, 'templates'. They weren't just beautiful; they were functioning areas of production, and, because they were so well managed, they eliminated the risk of uncontrolled fire.

The existence of infrastructure, houses, fences, outbuildings, and power lines complicates the adoption of a similar method, but does not prevent it. We just have to think differently about the country. And we have to investigate the ability of Aboriginal people to manage fire. Australian ignorance of Aboriginal culture allowed the myth that Tasmanian Aboriginal people did not know how to light fire to survive for two centuries, until Beth Gott scotched this in 2002. The paucity of archaeological research in most areas of Australia has allowed such fallacies to survive, and we must fund and encourage our scholars to search for real knowledge of Aboriginal fire management. That search, of course, should begin with discussions with Aboriginal Australians.

6

The Heavens, Language, and the Law

There was something going on in Australia. The explorers noticed it, and some wrote about it; but, as their primary purpose, and that of their sponsors, was to find land for European farming, many of their observations were allowed to slip from view.

Even so, as contemporary academics and researchers revisit the underpinning assumptions deployed in Aboriginal studies, many archaeologists and linguists have been led to examine the period of 'intensification' of food production and technology that many contend occurred around 4,000 to 5,000 years ago. While these new directions are encouraging, some cultural assumptions can still be seen at play.

At the 2013 history symposium at the Australian

National University, the respected linguist Michael Walsh cautioned that theories about the time of intensification and an associated language shift remain just theories, requiring a much greater level of testing.

Interpretation of Aboriginal society by archaeologists and anthropologists reflects, according to US anthropologist Aram Yengoyan, the reigning fashion in anthropological theory that often fails to acknowledge rites and ceremonies as religious and philosophical acts.[1] 'Not only does morality result from the Dreaming, but virtually all behaviour is an expression of a well-developed sense of moral conduct which provides the basis for all human imperatives.'[2]

If the culture of Aboriginal society is not given sufficient credence, it is easy to misinterpret the achievements. The economic foundations of traditional society were inseparable from the philosophic and religious beliefs, and to see the spiritual life as simply superstition and myth means that the practical advances in food production become invisible.

As discussed earlier, Darwinism was a profound driver of nineteenth- and twentieth-century thought and deed. It inspired amateur ethnographers such as Howitt to try anything to prove that the Aboriginals were a dwindling arm of the human family. Howitt had a more generous relationship with Aboriginal people than most, and employed them on his hop farm in Gippsland. However, he wrote to his sister, 'I arrive at the conclusion that the

Australian black is a "wildman" by nature and that you cannot "wash a blackamoor" … they have the minds of children and the bodies of adults.'[3]

The historian Bain Atwood commented that Howitt 'had quite different motivations from the Aborigines in the various activities which formed the basis of their relationship; that they had mutual interests was fortuitous rather than the result of any particularly benevolent attitude or behaviour on Howitt's part'.[4]

While some explorers referred to their Aboriginal guides by a single Christian name, Howitt referred to his as 'a native boy', 'black boy', or 'a black'.[5] Although he was scornful of Aboriginal culture, he was conscious of his fame as a black expert, and badgered the people on his farm to demonstrate initiation rites, even though they firmly resisted him. Eventually, he brought out sacred items in his collection, and pretended to the people that he had received them on initiation, which was purportedly how he had lost his front tooth. However, Howitt was never initiated, and was simply lying in order to advance his own theories about the superiority of Western man.

Howitt and others like him were operating under the encouragement of Australian institutions that believed they were there to supervise the disappearance, or at best, the submergence, of Aboriginal people and their culture.

If we are to attempt to understand Indigenous philosophy, it has to begin with the profound obligation to land. Deborah Bird-Rose comments:

> The state of the country, for instance, offers concrete evidence of the responsibility which the owners have been exercising. Responsibility is grave: there is no hiding in a conscious universe … the exercise of will in a situation where the choice to deny moral action is to turn one's back on the cosmos and ultimately on one's self.[6]

In supporting this view, Bird-Rose quotes the ethnographic theorist Bill Stanner: 'Aborigines have no gods, just or unjust, to adjudicate the world.' Therefore, he argued, all parts of the cosmos are conscious and must take responsibility for their actions.[7]

There is no separation between the sacred and non-sacred, as all actions are steeped in religious purpose. The mere act of burning grass has a specific story attached to it, a story that directs the need and benefits of the action in law. Stanner was convinced that all myths were archetypal, and 'enfold into some kind of oneness the notions of body, spirit, ghost, shadow, name, spirit site and totem'.

He reviewed post-war intellectual thought in Australia, and found that very little of the burgeoning self-conscious nationalism considered Aboriginal people to be part of the national fabric. Our intellectuals were looking at a 'view from a window which has been carefully placed to exclude a whole quadrant of the view'.[8]

This is the window through which our nation angles its view of Aboriginal Australia: the assumption of

Indigenous inferiority shackling theory to ideology.

The debate around Aboriginal and Torres Strait Island languages is hampered by fixed views on when people first arrived in Australia. Some linguists are arguing that languages were pushing north to south, inferring that they came from somewhere else. The problem would seem that recent archaeological discoveries are pushing back occupation dates of Australia with every new survey and would appear to threaten the date given to intensification. Many of the theorists pondering the time of agricultural intensification in Australia seem to ignore evidence of much earlier agricultural practices than their dates of 4,000 to 5,000 years ago.

Like the 'migrationists' of the twentieth century whose every idea was governed by the assumption of racial superiority, so too the 'intensificationists' seem to be influenced by the idea that humans *must* be on a perpetual trajectory of growth. Most ascendant civilisations, however, eventually run into dead ends, and current Western civilisation, and its obsession with growth, has been called into question by theorists such as Jared Diamond, who cautions that civilisations are destined for the fate of the Romans, Phoenicians, and Eygyptians.

If, however, you accept what the explorers said they saw, Aboriginal history is much more complex than our nation believes. And if you accept that to be true, the next thing to ask yourself is: How was that made possible? How was that system managed?

The subtle but comprehensive management of the land and its productivity is a worthy study in a nation intent on economic sustainability. The fact that it may cause a change in how we view land ownership and management is not a cause for panic. Fencing is one of the great differences between Aboriginal and European land use, but it is not impossible to imagine an agriculture without it.

When massive tractors were capable of drawing huge harvesters, it wasn't long before some neighbouring farms in the Wimmera region, in western Victoria, pulled down their fences so tractors could drive in straight lines. It was an economic imperative and created some early difficulties, but farmers adapted to the change because it was in their interest to do so. As Gammage says, 'Fences on the ground make fences in the mind.'[9] You have to alter your thinking to imagine the fence away.

We could do something similar in our application of fire. There may be early inconveniences, but, once the benefits are realised, we would adapt very smartly as long as we keep an open mind and don't confuse a logical change with the march of communism or a return to 'primitive' Aboriginal techniques.

There may not be a golden rule found in Aboriginal governance, but I suspect there are elements of agriculture, conservation, culture, and government that, having been tested against the nature of Aboriginal society for a minimum of 80,000 years, hold profitable messages for the nation.

Government

Arguing over whether the Aboriginal economy was a hunter-gatherer system or one of burgeoning agriculture is not the central issue. The crucial point is that we have never discussed it as a nation. The belief that Aboriginal people were 'mere' hunter-gatherers has been used as a political tool to justify dispossession. Every Land Rights application hinges on the idea that Aboriginal and Torres Strait Islander people did nothing more than collect available resources, and therefore had no managed interaction with the land; that is, the Indigenous population did not own or use the land.

If we look at the evidence presented to us by the explorers, and explain to our children that Aboriginal people *did* build houses, *did* build dams, *did* sow, irrigate, and till the land, *did* alter the course of rivers, *did* sew their clothes, and *did* construct a system of pan-continental government that generated peace and prosperity, it is likely we will admire and love our land all the more. Admiration and love are not sufficient in themselves, but they *are* the foundation of a more productive interaction with the continent.

Behaving as if the First Peoples were mere wanderers across the soil and knew nothing about how to grow and care for food resources is a piece of managerial pig-headedness. Smart business people rule nothing out, especially if the seeds of success are obvious.

The songlines of Aboriginal and Torres Strait Islander

people connected clans from one side of the country to another. The cultural, economic, genetic, and artistic conduits of the songlines brought goods, art, news, ideas, technology, and marriage partners to centres of exchange.

The Brewarrina fish traps were one such centre, the Lake Condah eel fishery another, Sturt's grain fields of the Warburton River region another, and Melbourne's Botanical Gardens were the point of dispatch for the great Dreaming corroborees brought from the Australian Alps by such important philosophers as Kuller Kullup.

We can trace the green stone axes from Victoria's Mount William quarry along the routes of that exchange; we can see elements of dance and music trading ideas back and forth across the nation. And, if we look, we might even find that indigenous plants flourishing in new homes were first brought to those regions by the hands of black traders.

If we accept that Aboriginal people were managing their landscape and economy across cultural and geographic boundaries, we need to consider how that co-operation was wrought without resort to the physical coercion and war common in other civilisations.

In all the archaeology and all the investigation done to date, there has been no time identified when those trade routes were used for wars of possession. The Grecian and Roman frescoes and ceramics feature war and torture as an element of dominion; but while individual acts of violence are depicted in Aboriginal art, there is no trace of imperial warfare. This absence demands respect, and the skills employed to bring about the longest lasting pan-continental stability that the world has known must be investigated, because they might become Australia's greatest export.

Behind the green bough of peace brought by ambassadors of distant clans, and the excitement of the trading market, there must have been an intellectual musculature, not just to forge that peace, but to maintain it.

Australian anthropologist Ian Keen says:

> The genius of Ancestral law was that people of a wide region could agree to a body of legitimate law without there being legislation, and in spite of the autonomy of individuals and kin groups ... Ancestral Law had a large discretionary component ... (and) certain ritual practices tended to induce in young people a disposition to conform to shared values and norms, and to defer to people in authority.[10]

In 1838, the missionary of Buntingdale in south-west Victoria, Francis Tuckfield, was alarmed by the violent disputes between Aboriginal clans, but came to realise that these only occurred when the people were forced into contact with their enemies, and his mission required just that. Travelling more widely in the country, and living and fishing with Aboriginal people of the Murray River, he was impressed by the peacefulness and respect underpinning the community. He became aware that the violence he'd witnessed at Buntingdale was a result of mission management, not the nature of the society.[11]

Dr Suzanne Davies has concluded that, in dealings with Aboriginals, equality may have been espoused, but the denial of the right to own land or give evidence in court removed Aboriginal people from any benefit of citizenship. The reality was that Europeans exerted 'control over Aborigines actions that they (Europeans) required … to gain access to the land'.[12]

There is no doubt that Aboriginal life was not one long dream of peace and harmony. Anger, bitterness, betrayal, revenge, and punishment were all common, but they were governed by strict rules. The violence was often punishment enshrined in the execution of lore, the tried and true systems of cultural and social and religious maintenance.

It's difficult to look at the decision-making processes involved in the creation of Aboriginal and Torres Strait Island government and not think of the word 'democracy'. Were the Elders elected? Not all who became old were included in the final decision-making processes; that authority was received following the complex trials of initiation.

To that extent, Elders became the equivalent of senior clergy, judges, and politicians. Their role was codified by levels of initiation that elevated them to a position where they could influence particular areas of policy. Their election to that position was gradual and convoluted through the initiation process, but they didn't assume that position as a result of force or inheritance. They earned the respect of their fellows.

All other processes of delivering justice, protecting the peace, managing social roles and dividing up the land's wealth were defined by ancestral law, and interpreted by those chosen as the senior Elders. Of all the systems humans have devised to manage their lives on earth, Aboriginal government looks most like the democratic model.

For a model to remain sufficiently coherent and flexible, and appeal to such a large population spread over such a large area and such great time, requires our serious regard. It must have appealed to the vast majority of Aboriginal Australians as having an internal logic and fairness; otherwise it could not have survived.

The social cohesion arising out of this system of government allowed people to co-operate in all aspects of food procurement. Large numbers of people could gather to contribute the massive labour involved in the construction of dams, fish traps, and houses, and in the preparation and maintenance of croplands. Without political stability, activities extending beyond language and cultural boundaries would have been impossible.

An indication of how Aboriginal cultures emphasised peace and stability is reflected in a central explanation of Yuin culture on Gulaga Mountain, near Tilba Tilba on the south coast of New South Wales. At a gallery of massive rock formations on the girdle of the mountain's flank, you are asked to pause at the Healing rock and consider the welfare of the unborn child, the ill, the troubled, and then you are introduced to Nyaardi and Tunku, the first woman and man. Nyaardi is twice as tall as the mere male, Tunku. In between them are the gifts given to them by the creator, Dharama, the tree and the stone. Everything they will ever need can be derived from those two gifts. Next we see the pregnant Nyaardi, and you're asked to gently place your hand on her belly; later, you see her carrying the child on her back, and you pass the three great stones of human existence: past, present, and future. The present is the larger stone. Past and future must be considered, but the most important target of your thought is to do with the now. There is the great ark of life, the birth canal of woman, and the rock where all humans learn and consider their path.

At this point, you leave the chapters of Yuin law and realise that you have not seen a weapon, not heard one story of a vanquished foe, but instead the centrality of women to human life and the respect that woman, the mother, must be shown. Through the mother is the saying Yuin men use in formal situations: we all arise from the Mother. That is the lore.

Anyone can visit that mountain, although it is best to climb it barefoot and silent, and to take care not to harm either animal or plant. We are responsible for the world's health; not it for ours.

Touring the world's churches, art galleries, castles, and museums cannot prepare you for this experience. The severed heads in the galleries of Venice and Milano, the molten lead poured into the mouths of those who disagreed with French kings, the drawings of wretched poverty, the frescoes of war, the murder of infants are common themes, and European and Asian cultures are suffused with this violence and war. Yet here on a green mountain overlooking the Pacific, a different world was imagined. And not just imagined, but lived. (Ngaran Ngaran Culture Centre, Narooma, and the Gulaga Mountain Management Committee run tours of the mountain.) Eighty thousand years is forever. Aboriginal people say we have always been here, while many Australians claim that, having arrived from Africa, Aborigines are just migrants, boat people, like everyone else.

In recent years, linguists have been attempting to find the proto-Australian, or root, language of Australia in order to better understand the human history of the continent. Of course, the assumption is that Aboriginal people *had* to come from somewhere else. In terms of language, culture, and religion, no system has seen such prolonged and stable development, but still the theorists' subconscious searches for an explanation of the Australian aberration.

It is fascinating to contemplate the human evolutionary ascent, but it must also stimulate our minds to wonder at the development of the great Australian peace. Isolation and depravation were considered the reason Aboriginal people did not 'advance' like Europeans, but it is also true the idea to pour boiling oil on enemies seems not to have occurred to anyone in Australia.

It is almost certain that isolation had a huge impact on the trajectory of the development of the Australian cultures, but that isolation may also have allowed time for peace to be forged. Or was it a factor of the old continent itself? The relatively low fertility of Australia and the general erosion of the land may have provoked a different human response. Who knows?

Language

To better understand the development of Australian languages, grammar analyses have been undertaken in the light of recent archaeological discoveries. Patrick McConvell, Nick Evans, and Isobel McBryde all have interesting theories that there was a sudden southward push of language about 5,000 years ago, presuming some kind of influence from southern Asia.

Other linguists, such as Terry Crowley, believe that comparative linguistics is the way to go, but the prototype language has not been found in the area where it is presumed Aboriginal people arrived. Crowley admits that Australian languages are probably 40,000 to 60,000 years old, but even at 10,000 years they would be older than most other world languages.

In line with other linguists, Crowley theorises a sudden southern shift of language during the period of technological intensification about 4,000 years ago. The evidence supplied to support this theory is, however, full of perhapses and maybes.

Some kind of re-invasion of Australia provoking the sudden technological change is not supported by genetics. If a northern group stepped ashore in Australia and drove people south ahead of them, you would expect the northern group to show more diverse genetic characteristics than other parts of Australia, but that has been proven not to be the case.

Others have tried to find evidence of late invasion from the north in the names applied to more recent tools, but the evidence to support this theory is at best sketchy.

Neville White and Isobel McBryde came to the conclusion that the spread of languages was not driven by refined tool technology, but rather by spiritual and social change. According to McConvell and Evans:

> [White] aims to explain language spread without postulating conquests (ethnologically unlikely in the Australian setting), or major migrations into unpopulated areas; the latter is implausible given the antiquity of evidence of occupation, and incompatible with the genetic evidence that Pama-Nyungan speakers do not exhibit less genetic diversity than other parts of Australia, except in the Central Desert region.[13]

(Pama Nyungan languages form the bulk of Australian languages, with the non-Pama Nyungan being centred mainly in the Kimberley and Arnhem Land.)

Rhys Jones and Nicholas Evans also see spiritual culture as an important determinant:

> [A]ttempts to see linguistic expansion in the simplistic material view of conquest and intensification run up against a linguistic problem: the apparent unreconstructability of a good number of wooden artefact terms. We have therefore proposed an alternative scenario in which new technology was spread in association with a particular set of rituals, with initiates being inducted into Pama-Nyungan as they learnt new ceremonies and new tool-making techniques, and linguistic expansion being driven by ceremonial prestige and changed patterns of spouse export as Pama-Nyungans demanded payment in wives for their sons, leading to export of Pama-Nyungans to new households. Underlying the social innovation of new ceremonies and wider alliances were advances in food technology that allowed large gatherings to be fed for reasonably long periods.[14]

Jones and Evans conclude with this summary: 'As two nearly incurable materialists we therefore dedicate the chapter to the many Aboriginal people who have again and again taught us the priority of the spiritual and social dimensions of technology.'[15] They illustrate that idea by talking about the spiritual weight given to the raw material for tool production, and how weapons and tools derived from stone and wood are loaded with moral and spiritual obligation and significance, all of which is reflected in language.

It's highly theoretical, but Evans and Jones are highlighting a dramatic change of social and cultural activity rendered not by conquest and invasion, but by the sharing of cultural knowledge and development. Gammage also argues that, 'Local populations remained stable enough to stay within country. There were no population driven conquests.'[16]

Whether the spread of language occurred exactly as Jones and Evans have postulated requires more study, but the fact that everything points to a significant cultural change without dramatic violence or displacement is extraordinary. The political processes required to manage such a transition had to be profound, and there must have been persuasive social forces at work in order for them to succeed over such a vast time.

Josephine Flood takes theorists such as Peter Hiscock to task, arguing that their 'change equals progress' assumption is influenced by a politically correct Western view of cultural advancement. Flood believes Aboriginal culture 'was surprisingly stable and change was relatively slow'.[17]

Michael Archer also has an interesting view of 'progress':

> European colonists cleared or damaged bush because they did not value it and introduced to more than sixty-five per cent of the continent mono-cultures of non-Australian species they did value ... it is our southern Eurasian ancestors ... who are actually nomads because we overpopulate ... damage land in the process, then wage wars on neighbours to take their land in order to continue to over-populate, and on it goes.[18]

Further, he says, environmental destruction in western New South Wales, where the primary degradation of land and extinction of mammals occurred, was due to the introduction of sheep.

Rolls elaborates on that point in *A Million Wild Acres*. I was led to Rolls' book because it turns up as one of the three references that appear if you google Brewarrina fish traps. I'd read only excerpts from the book when it was released, but was impressed by its sympathetic approach to the Australian landscape.

Reviewers claimed it was unusually sensitive to Aboriginal Australians. There are a few references to the displacement of Aboriginal owners of the Pillaga region, and some allusions to the atrocities employed to displace them, but the book is essentially an analysis of white settlement. The dearth of material examining the Aboriginal economy at the time of publication meant that the mere mention of Aboriginals could make the book seem radical to the public view.

Archer took the idea further. He compared the expansionist ethos with the more conservative Aboriginal practices. That the sustainability of Aboriginal economies could be linked to more sustainable theories of government is an electrifying idea, because it has the potential to influence decisions on agriculture, population targets, water allocation, and environmental protection. It's not touchy-feely wise blackfellow versus the destructive imperialist whitefellow; it strikes instead at the heart of conservative economic practice and the evolution of species.

Some say the idea that the world's trajectory is driven by conquest followed by innovation and intensification is satisfying to the Western mind because of our psychological dependence on our imperialist history. But if we give consideration to the idea that change can be generated by the spirit, and through that by political action, the stability of Australian Aboriginal and Torres Strait Islander culture might be more readily explained.

As all the theories are very tentative and untested, we should be wary of locking ourselves into the assumption that everything is driven by superior Western minds and tools on an inexorable march of conquest, as if that is the only way the human species might evolve. Ian Keen's study of how economies were embedded in the prevailing kinship and cosmological systems of particular groups has the potential to provide insights into how Aboriginal societies worked.

How language changed over time and where such change originated is part of a new and tentative science. Peter Hiscock warns that some current studies trivialise Aboriginal history, and miss 'the dramatic and noteworthy' points in its cultural and social development.

The whole debate is speculative, and swings on the view held about when Aboriginal people first arrived in Australia. Those with the progressionist ideology of conquerors instigating cultural and technological advance will be challenged by the changing perceptions attaching to the origins of the human family.

McConvell argues that Aboriginal mythology did not arise as a fundamental block, but in stages by different founding ancestors. Rock-art expert George Chaloupka refers to western Arnhem Land, where one ancestral figure populated the land with people, plants, and animals, and designated which languages were spoken, while later ancestor figures instructed the people how to structure human society.

From this it is assumed that languages could be spread without conquest and the surges of technology characteristic of linguistic change in other parts of the world. The argument proposed by Peter Bellwood in *First Migrants* was that you would only see a coherent and pervasive language spread if it happened slowly. If it happened quickly, as is the case with invasion by conquerors, language differences would persist in family groups.

The argument of invading conquerors as instigators of change is not supported by the fact that the Murray Valley region had been diversifying culturally and technologically for at least 13,000 years; and that, after 10,000 years of isolation, Tasmanian Aboriginals remained genetically and culturally similar to mainland peoples.

McConvell, in studying kinship terms, found a consistency across most of the country, and believes that the majority of these terms were inherited from the proto-language. Language study is showing us that social stability allowed a unique human response to food procurement. The science is dense and contradictory, but doesn't alter the fact that Australia is different, and that now is the time to celebrate and explore the difference.

Trade and economy

One of the central tenets of trading was the sharing of resources. The great bunya bunya nut harvests enabled huge gatherings of people to enjoy trade and cultural relationships with sufficient food to sustain all participants for long periods. In the south, the phenomena of moth collection provided another opportunity for trade and cultural purposes.

In both cases, it would have been possible for the territorial custodians of the resource to keep it for themselves and simply stockpile more and more food, and gain more and more trade credits. But they chose to share the resource, actively pursuing the opportunity to attract other clans into their country for the purposes of cultural and social exchange. The resource was more than a commodity; it was a civilising glue.

One of the greatest differences between the culture of Aboriginal Australia and that of mainstream Australia is the concept of land. The same liberal philosophy that did so much to abolish slavery also promoted the rights of the individual, and that meant individual ownership of land. Tony Barta comments:

> [T]he small band of men and women who had created the first modern political movement to defeat slavery in the British Empire met their match on the South African veldt and the pastures of Australia Felix. They understood only too well the relationship between the appropriation of land and the loss of Aboriginal lives but found their power to intervene more circumscribed by distance and a different spread of interests ... the mass appeal of property and opportunity was only beginning to gather pace. In Australia ... it was plain what that meant for indigenous peoples in its path.
>
> The European project was about taking possession of land, asserting dominion over nature, despoiling newly 'discovered' country to create a familiar civilisation.[19]

Aboriginal Australian law insisted that the land was held in common and that people were the mere temporal custodians. Individuals were responsible for particular trees, rivers, lakes, and stretches of land, but only so these could be delivered forward to the next generation in accordance with law. Individuals and families might be said to own a particular fish trap or crop, but they worked it in co-operation with the surrounding clans.

The combination of joint ownership, or belonging to land, and the co-operative utilisation of crops and animals adapted to Australian conditions, meant that fencing was rare and impermanent. Battues could be brought into operation when required in the same way that fish-trap fences could be opened or closed according to the needs of both people and animals; but, significantly, they did not impede access to the landscape.

The system in operation could be considered a jigsaw mutualism. People had rights and responsibilities for particular pieces of the jigsaw, but they were constrained to operate that piece so that it added to rather than detracted from the pieces of their neighbours and the epic integrity of the land.

The piece of the tree or stream or land that a group retained responsibility for bled into country so distant that they might never visit that country. They had to imagine how the whole picture looked, and they had absolute confidence in the coherence of the accretive construction of their law over thousands of years, and knew that the jigsaw would make sense and that their responsibility was to ensure it continued to make sense.

People fishing a family section of the great Brewarrina fish trap knew that they could take fish from that trap, but they had to ensure it was done in such a way that it did not impede enjoyment of the amenity by countrymen and countrywomen they would never see.

The religious, social, and governmental rules were forged and entwined in a mandala that had to be imagined in the soul.

Bill Stanner said, 'Our own intellectual history is not an absolute standard by which to judge others. The worst imperialisms are those of preconception.'[20]

If you analyse any element of human life on the assumption that Western thought and Christianity are the unassailable pinnacles of human development, you are bound to find the unbelievers inadequate, and that means you close yourself to the depth and subtlety of any other spiritual manifestation.

Stanner claimed that, at its core, Aboriginal belief is about abidingness, that:

> [T]he philosophy of assent, the glove, fits the hand of actual custom almost to perfection, and the forms of social life, the art, the ritual, and much else take on a wonderful symmetry … the notion of aboriginal [Stanner always used a lowercase 'a' for Aboriginal] life as always preoccupied with the risk of starvation,

> as always a hair's breadth from disaster, is as great a caricature as Hobbes' notion of savage life as 'poor, nasty, brutish and short.' The best corrective of any such notion is to spend a few nights in an aboriginal camp, and experience directly the unique joy in life which can be attained by a people of few wants, an other-worldly cast of mind, and a simple scheme of life which so shapes a day that it ends with communal singing and dancing in the firelight … its principle and its ethos are variations on a single theme-continuity, constancy, balance, symmetry, regularity …
>
> One of the most striking things is that there are no great conflicts over power, no great contests for place and office. This single fact explains much else, because it rules out so much that is destructive of stability … There are no wars of invasion to seize territory. They do not enslave each other. There is no master-servant relation. There is no class division. There is no property or income inequality. The result is a homeostasis, far-reaching and stable.[21]

Charles Sturt observed that, '[I]t is a remarkable fact that we seldom or ever saw weapons in the hands of the natives'.[22] The most common weapon doubled as a paddle for a canoe, and was so heavy as to be generally unsuitable for offence.

Sturt wrote in his journal:

> The character and spirit of these people, is entirely misunderstood and undervalued by the learned in England, and the degraded position in the scale of the human species into which they have been put, has, I feel assured, been in consequence of the little intercourse that had taken place between the first navigators and the aborigines… I have seen them in a variety of circumstances — have passed tribe after tribe under the protection of envoys — have come suddenly upon them in a state of uncontrolled freedom — have visited them in their huts — have mixed with them in their camps, and have seen them in their intercourse with Europeans, and I am, in candour, obliged to confess that the most unfavourable light in which I have seen them, has been when mixed up with Europeans.[23]

Sturt bemoaned the inevitable conflict with Europeans as more and more Aboriginal land is traduced, and was certain that the arrival of the European was an invasion rather than a religious mission: 'I have to regret that the progress of civilized man into an uncivilized region is almost invariably attended with misfortune to its original inhabitants.'[24]

He acknowledged the poverty and depravity visited on the Aboriginal population when in the company of Europeans, but his first instinct was not to consider restricting the area usurped by Europeans; rather, it was to take Aboriginal children away from their families: 'the only remedy involves … a complete separation of the child from its parents … [no] good will result from the utmost perseverance of philanthropy … until the children are kept in such total ignorance of their fore fathers, as to look upon them as Europeans do, with astonishment and sympathy'.[25] Mitchell had a similar idea, and took an Aboriginal child, but when he returned to England the child was an impediment, and he abandoned it.

So many Europeans thought their Christian task was to smooth the pillow of the dying race. Aboriginal people consider that as just a smokescreen behind which to *steal* the pillow of the *living* race.

Mitchell talked in sorrow of the demise of Aboriginal Australia:

> These unfortunate creatures could no longer enjoy their solitary freedom; for the dominion of the white man surrounded them … hemmed in by the power of the white population, and deprived of the liberty they formerly enjoyed of wandering at will through their native wilds, were compelled to seek a precarious shelter among the close thickets and rocky fastnesses which afforded them a temporary home.[26]

But despite this compassion, Mitchell writes, a mere two paragraphs later, 'We again (on the Hunter) find some soil fit for cultivation, and the whole of it has been taken up by farms. But the pasturage afforded by the numerous valleys on this side of the mountains ... is more profitable to the owners of the farms.'[27] At one moment he expresses sorrow for the losses of the Aboriginal population, but within a page he's extolling the value of the lands forcibly taken from them.

On his previous explorations, Mitchell had seen the use that Aboriginal people made of their lands, although some of the food-production techniques were too discreet to capture his attention or understanding, but then opined mildly about the future of Australian farming, as if Aboriginal food production had never existed. He looked down the valley at the sheds and houses of the settlers, where smoke dwindled from the chimney, and squares of amber light glowed at windows, and revelled in the domesticity. Only a year before he was envying the warmth, cheer, and domesticity of the Aboriginal village, but now he prided himself on opening up this land to his own race.

He's a good man, Mitchell, but he shared the ambition of every British man in the colony — land:

> We crossed ... a rich plain where the grass was remarkably good ... We were delighted with so favourable a country for extending our journey [and expected] it might open out a field of useful discovery.[28]

Later, Mitchell exulted on looking over the plains of his Australia Felix in Victoria:

> A land so inviting, and still without inhabitants! As I stood, the first European intruder on the supreme solitude of these verdant plains, as yet untouched by flocks or herds; I felt conscious of being the harbinger of mighty changes; and that our steps would soon be followed by the men and the animals for which it seemed to have been prepared.[29]

The confidence of Mitchell in assuming that the land had been waiting for Europeans and their animals is at the heart of European intellectual arrogance. Europeans failed to see the symbiotic nature of the Aboriginal spirituality and economy. Of the hundreds of Aboriginal legends that highlight the rules of relating to the country when procuring food, perhaps a few might give an indication of that relationship.

The story of the killer whales and the Yuin people around the south coast of New South Wales quoted earlier has many curious and fascinating layers, apart from mere food provision. The nineteenth-century whale industry at Twofold Bay manipulated that relationship, and Aboriginal men and women co-operated with European whalers in the fishing operation.

Beryl Cruse, Liddy Stewart, and Sue Norman give a wonderful overview of the Yuin relationship with the sea in their book, *Mutton Fish*. They relate the traditional whaling story as recorded by ethnographer R.H. Mathews:

> When the natives see a whale being chased by killer whales one of the old men pretends to be lame and frail and lights some fires a little distance apart on the shore and walks between them pretending to be lame and helpless to excite the compassion of the killer whales and the man calls on the killers to bring the whale ashore. When the injured whale drifts in to shore the other men come out of hiding to kill the whale and call on neighbouring tribes to join the feast.[30]

This relationship had been in operation long before Europeans arrived, and continued, modified by the demands for profit, for many years afterward until a disgruntled European whaler shot the lead killer whale. That was the last time the cetaceans co-operated with men. The age-old relationship of reciprocity had been destroyed.

The most intriguing thing about the relationship between humans and killer whales was that it combined an economic imperative with a fundamental spiritual connection with the animal world. The totemic system of Aboriginal and Torres Strait Islander Australia insists on the interconnectedness and spiritual equality of all things. The Yuin believe that after death they return as killer whales. As an indication of this pervasive belief, massive land sculptures featuring the whale were seen as far as 100 kilometres inland.

The intimate co-operation between people and

cetaceans was remarked upon by many early colonists. Foster Fyans saw Aboriginals fishing at Geelong in partnership with dolphins that drove the fish into the shore to facilitate the Aboriginal catch, and similar relationships were reported at Moreton Bay, Monkey Mia, and many other Australian beaches.

John Blay, in his survey of old Aboriginal pathways in the south coast of New South Wales, links those ways to the ceremonies for the harvest of Bogong moths in the mountains, and the whale feasts on the coast. He asks his readers:

> Where else in the world were there gatherings like those for the Bogongs? Where else anything like the association with Orcas and whale hunting in Twofold Bay? These will deservedly come to rank amongst the great stories of Australian culture, as cornerstones of what and who we Australians are. These stories exemplify sharing the abundance.[31]

In Aboriginal life, the spirit and the corporeal world are wedded; but in European society, the economy operates independently of the spirit, and, as modern examples illustrate, almost in defiance of the religious moral code. The financial crash of 2009 and the oil spill in the Gulf of Mexico in 2010 occurred because the Christian morality of most participants had been excluded from their business dealings. In the case of the oil spill, it highlighted the dominion that Christians believe they hold over the earth.

The ability of the planet to survive such cavalier business codes in the financial world or the safety provisions for drilling operations of global oil companies is being sorely tested. The old man and the killer whale remains a salutary lesson for us today.

The stories above are strictly moral in outlook, and emphasise sharing and truthfulness, but even in this form are probably sheltering critical information from the uninitiated. Other stories spoke about the night sky, where mythical beings played out their eternal roles; some of the stories related methods of weather forecasting, but in general they were stories about the law of the land.

7

An Australian Agricultural Revolution

One of the most fundamental differences between Aboriginal and non-Aboriginal people is the understanding of the relationship between people and land. Earth is the mother. Aboriginal people are born of the earth, and individuals within the clan had responsibilities for particular streams, grasslands, trees, crops, animals, and even seasons. The life of the clan was devoted to continuance.

The intensification of resource use, language development, and social organisation were in the curve of great change prior to the colonial period, because Aboriginal and Torres Strait Islander people were on the same cognitive trajectory as the rest of the human family, albeit in a different stream and a unique channel in that stream.

Perhaps the most significant difference was the attitude to land ownership and resource use. Instead of operating privately owned small holdings, clans were co-operating to prepare large areas of land for production with burning and tilling methods. There was an underlying conservatism in this approach, a concern for people they might never meet, and a respect for the prey species embedded in the spiritual and cultural fibre.

If we could reform our view of how Aboriginal people were managing the national economy prior to colonisation, it might lead us to reform the ways we currently use resources and care for the land. Imagine turning our focus to the exploitation of meat-producing animals indigenous to the country. Imagine freeing ourselves from the overuse of superphosphates, herbicides, and drenches, and freeing ourselves from the need of fences, and instead experimenting with grazing indigenous animals and growing indigenous crops.

Superphosphate was introduced when vast quantities were found on Nauru in the 1930s, and it was subsequently sold to farmers as the boon fertiliser. Some agricultural scientists are now questioning its role in the salinisation of vast areas, and the poisoning of water supplies — it is accepted that fluorine leaches from superphosphate fertilisers. Super also kills many of the creatures and microbes that actively assist the growth of plants. In analysing our impact on the land, we need to review the use of this fertiliser. Farmers are among the many who have been calling for this reanalysis.

Farmers are adaptive entrepreneurs, and when the public demand switches from red-wine grapes to white-wine grapes, or from beetroot to olives, the farmers respond. Agriculturalists will change, but will the consumers?

Our agricultural ministries and research institutes have begun looking at some of the Aboriginal food products, but tend to concentrate on the most popular, trendy foods, such as lemon myrtle, bush tomato, and bush raisin. Schoolchildren are taught that witchetty grubs were a major food source, almost as if there is a deliberate attempt by educationalists to emphasise the gross and primitive.

Imagine, instead, re-educating the nation and utilising the two major crops of Aboriginal Australia: yams (as well as other root vegetables) and grains. All of these plants were domesticated by Aboriginal people, and these are the plants that offer the most exciting prospects for farming today.

Many plants have received too little attention. Kangaroo grass, *Themeda triandra*, is intolerant of overgrazing, and its seed yield per acre does not compare well to wheat and rice, but it may be the perfect plant for those marginal dryland farms where grain and sheep farms have been abandoned. The quality of flour from these grains appears little known; but if this is one of the grasses harvested to make the cakes that Sturt professed were the best he had ever tasted, perhaps we need to look more closely.

The barley *Microleaena stipoides* is thought by some scientists to be a suitable plant for commercial cultivation, but agriculturalists need support in trialling the sowing, harvesting, storing, and marketing techniques suitable for this plant.[1] Davies, Waugh, and Lefroy studied this plant's potential, and whilst its yield is not as great as an annual grain, they considered that its utility as both a grain and a grazing crop would make it economically viable. As mentioned earlier, the retention of carbon in the roots of perennial plants will prove to be the clincher in encouraging a shift toward perennials.

The yam daisy, *Microsceris lanceolata*, would seem another logical commercial crop likely to prove attractive to our food-conscious society. Following a year when an Aboriginal working group analysed Aboriginal living sites at Mallacoota in Victoria, several Aboriginal and non-Aboriginal members embarked on a program to grow yam daisy from seed. The trial is still in progress, and in the spring of 2012 we harvested our first seed and replanted most of it in the autumn of 2013. Various soil types and growing conditions have been trialled in order to increase our knowledge of the plant. Soon we will be able to sell seed and spread knowledge of the plant across a variety of gardeners and soil types.

Already, we have found that the plants grown from our own seed are more robust and productive than in the first season, suggesting that the plant has adapted to and responds to the processes of cultivation. The plants prosper when the soil around them is tilled and older plants develop larger basal leaves that lie flat on the ground and help prevent weeds. We have incorporated moss and bulbine lilies into our plots, and we're looking at kangaroo grass, which was remarked by many early observers as often growing up through crops of yam. The interplay of different edible plants was no accident, as we have to try and understand the reasoning and the benefits of these planting relationships.

Our aim is for one, or a group, of the young local Aboriginal people to turn the results of this investigation into a profitable industry.

One of the growers, Annette Peisley, has measured the energy quotients of various fruits and tubers using the Brix index. 'Potato has a Brix reading of around 5–6° therefore in terms of fructans (carbohydrate level and accepting the limitations of the Brix method) a 100g sample of *Microseris lanceolata* tubers would provide 3–4 times the energy level of a 100g potato.'[2]

Aboriginal people would have required a smaller volume of tubers to attain the same energy quotient, and that would have allowed for more efficient storage and transport.

Further investigation of the properties of the yam may lead to greater knowledge of its potential as a commercial crop.

It is hoped that when the yam is accepted as a commercial product, Aboriginal people will be invited to take part in the science and share the new prosperity. Aboriginal groups are already attempting to acquire land to conduct field trials in East Gippsland, and positive government intervention could be of immediate and practical assistance.

Left: Murnong. Right: Harvester tuber

(Lyn Harwood)

Gurandgi Munji (a Yuin company on the New South Wales south coast) is planning harvests of a number of grains, and early trials of flour production have had spectacular results. The flavour and aroma of the breads cooked so far have hinted that we are on the track of Sturt and Mitchell's light and sweet 'cakes'.

The Mutti Mutti, Latji Latji, and Barkinji women and children combined forces with Japanese artist Yutaka Kobayshi in 2015 to harvest and extract grain from native millet. The subsequent baking of the bread in ovens on the sands of Lake Mungo was a resounding success, and an important moment in our history. The potential for Aboriginal Australians to make a living from the traditionally domesticated plants is an exciting prospect for the country.

Left: Panicum bread baked at Lake Mungo, 2016. Right: Kangaroo grass

(Jeanette Hope; Lyn Harwood)

The same applies to the fishing industry. The first European settlers dismissed the abalone as mutton fish.

As we have seen, as soon as Asian demand made it a valuable commodity, Aboriginal people were locked out of its harvest. Even so, East Gippsland Aboriginal communities are trying to encourage the government to include them in new plans to conserve fish stocks, because it was Aboriginal people who conserved those stocks by intelligent harvesting and quota limits for millennia. There was marine abundance when Europeans arrived; but, 200 years later, all commercial target species are threatened, and some have virtually disappeared. Inclusion of Aboriginal people in the allocation of licences would seem an expedient economy.

Accepting the full history of the country has the benefit of discovering a whole new level of knowledge about sustainable harvests. Change may be required, but it does not lead to a preference for wilderness or the withdrawal of productive lands. New ideas and new methods will arise out of the very oldest land-use practices.

The country may eat less meat in the future, but we will always eat some. Harvesting kangaroos and wallabies will not endanger the population of macropods, but instead guarantee their protection. We just have to accept the fact that if we are going to source protein in the form of animal flesh, it would make sense to use the animals best adapted to our soils and climate, those that do least damage to our soil, and make least demands on our dwindling water supplies.

Animal rights and welfare groups quite rightly monitor the production methods of farms and the treatment of domestic animals, but the national abhorrence for the consumption of native fauna is threatening our soils and water supplies. Utilising these animals does not mean they will never again be 'seen in the wild'; rather, it guarantees that they will, whereas our current methods are seeing mass extinctions of animals adapted to an environment previously managed and shaped by Aboriginal Australia.

8

Accepting History and Creating the Future

Gavin Menzies, in his book *1421: the year China discovered the world*, pondered the prospect of the Chinese continuing their mapping of the world after 1423. He argued that the Chinese had already reached Australia and had begun to trade with Aboriginal and Torres Strait Islander people.

Just as this social and commercial trade had begun, however, lightning and fire destroyed the Forbidden City, and Emperor Zhu Di, already under political pressure, saw this as a sign that the gods were against him. His health and fortunes suffered a decline, and his political opponents dismantled the foreign policy of incorporating the world into the Chinese system of trade. Chinese naval exploration ceased abruptly.

Under Zhu Di's administration, the Chinese received envoys from countries they visited, treating them like royalty in China, and then returned them to their homelands, showering them with gifts as a way of cementing trade ties. There is a theory that some northern Australian Aboriginals visited China under this scheme when the bêche-de-mer trade was being forged.

Menzies' view of what happened instead provides a bleak comparison between European and Chinese foreign policies: 'Instead of the cultured Chinese, instructed to "treat people with kindness", it was the cruel, almost barbaric Christians who were the colonisers. Francisco Pizzaro gained Peru from the Incas by massacring five thousand Indians in cold blood. Today he would be considered a war criminal.'[1]

In effect, the Portuguese used Chinese cartography to show them the way to the East. Then they stole the spice trade, which the Indians and Chinese had spent centuries building. Anyone who might stop them was mown down. When fifteenth-century Portuguese explorer Vasco da Gama reached Calicut, he told his men to parade Indian prisoners, then to hack off their hands, ears, and noses.

Invaders like to kill the original owners of the soil they intend to plunder; but, even better than that, they like to humiliate them. Once that hard work is over, their grandsons re-write the history of the re-named land and paint their grandfathers as benevolent visionaries.

In describing how nations can insulate themselves

from the facts of history, Menzies noted that, 'American and European historians had managed to persuade the world ... that Columbus had discovered America and Cook Australia.'[2]

This fabrication is not unique; the history of colonialism is dense with examples. Mention has previously been made of gigantic megaliths at Mount Elephant. There are some stone arrangements in the area, but they are similar to other ceremonial structures elsewhere in Victoria. Colonists recorded inflated accounts of the structures, and then informed the press that only Europeans could have built them. They supported their assertions with an engraving, which purported to show massive Stonehenge-like plinths in association with rudimentary dwellings of debased natives. The Aboriginal hut is drawn as a simple canopy supported on one end by a string attached to a spear. *See*, the argument went, *these people could not have built these monoliths; therefore, they are a later, debased race.*

Bogus illustration of Mount Elephant plinths

(Public domain)

According to McNiven and Russell, such deceits 'helped European colonists legitimize their right to inherit the Australian continent. European colonization literally became a process of (re)possession of a lost domain of their heritage'.[3]

The urge to legitimise occupation has been compared by them to the warping of history and archaeology by Nazis to justify the extermination of the Jews. The workings of the European mind when looking for ways of justifying barbarity creates an era of myth and denial.

There were other colonists from other continents, but it was Europeans who attempted to dominate the world, sometimes by dominating each other.

In trying to explain the killing of 14 million at the hands of Soviets and Nazis over twelve years in a relatively small area covering the Ukraine, part of Poland, Czechoslovakia, and Romania, Timothy Snyder, in his book *Bloodlands*, discusses the motivations of the coloniser: 'In colonization, ideology interacts with economics; in administration, it interacts with opportunism and fear.'[4] He was explaining how someone else's idea could enlist otherwise sceptical people to carry out the detail of that idea.

There is a photograph of early explorer and pastoralist Angus McMillan sitting with two Aboriginal men.[5] The image has always disturbed me. The man who altered his journals to imply that he was the first white man into Gippsland is holding the hand of the man on his right, who stares into the camera with a look of absolute fear. The other man stares with miserable resignation. McMillan's posture seems slightly strange until you realise he is sitting forward of the other two, and the plane of his thighs suggests he is sitting on a cushion so as to appear the taller of the three. McMillan's eyes seem to interrogate the photographer to assess if this staged piece has been sufficiently convincing.

The men in the photograph with McMillan are Big Johnnie Cabonne and Jemmy Gebber. McMillan not only altered the record of his travels to enhance his reputation as explorer, but introduced photographs to suggest he was a friend of the blacks, even though he and his men murdered hundreds.

Big Johnnie Cabonne, McMillan, and Jemmy Gebber

(State Library of Victoria)

Both Cabonne and Gebber were forced into McMillan's service to guide him around Gippsland. At one point, McMillan claims he found Gebber with a tomahawk upraised in the act of murdering him, but this is probably an excuse for McMillan's need to hold a pistol to Gebber's head to make him lead him into Gippsland.

Gebber's contemporary family have long memories, and claim that McMillan later killed Gebber, just as other Gippsland founding fathers murdered their guides upon being delivered to their El Dorado. It is interesting that Gebber freely led members of the Reed and Sellers families to land around the Delegate area. Perhaps, in despair of ridding his country of white men, Gebber was trying to people his country with men of gentler disposition. Accepting the best white man available is the final stage in the colonisation of Australia.

Some settlers were considerate of Aboriginal people, and grateful for their assistance and knowledge. Henry Sellers had close Aboriginal friends throughout his life, and the Reed family continued to employ Aboriginal men and women right up to the 1930s.

But not all were convinced of the humanity of the original owners of the land. Within months of a large massacre occurring in East Gippsland, where all but one child was killed, settlers began talking airily and in apparent bemusement about the disappearance of the blacks. Subsequent histories repeat this cauterisation of memory, despite the massacre being acknowledged by the first squatters of the district.

Ernest Favenc, an explorer who took violent exception to the existence of Aboriginal people, wrote a novel in which they were not the 'real' occupiers of the soil prior to European arrival, but were preceded by a superior ancient civilisation. In *The Secret of the Australian Desert*, this lost race is destroyed by a volcano, leaving the possession of the Australian continent to the Europeans. So, in one sweep, Aboriginal ownership is obliterated, any suspicious-looking structures are explained, and the continent falls bloodlessly into European ownership. This was his dream.

Historian Melissa Bellanta analyses this literary phenomenon, which she refers to as the Lemurian theory, after G.F. Scott's 1898 novel, *The Last Lemurian*. Australians were looking for a more worthy history on which to base their occupation of the land. Bellanta explains that, '[T]he figure of a once-grand civilisation in the interior endowed the Australia imagined in these works with a "mythical history" more worthy of its stature than that of the scattered nomads.'[6]

In the excision of unpalatable parts of our history (the illegal occupation of land and the slaughter of the occupants, for instance), we have lost elements we never knew existed. Those elements — such as the crops, houses, irrigation systems, and fisheries — may hold keys to future prosperity.

But first we need to understand how we came by the soil. Tim Flannery, in *Here on Earth*, compares the

Darwinian and Wallacian points of view, and those who have considered the evolutionary question in the last four decades. Peter Ward, for instance, suggested that the reliance on the survival of the fittest created a Medean outlook where ruthless competition could lead to the destruction of resources and populations. James Lovelock, on the other hand, posited the Gaia theory, whereby humans worked with a degree of cooperation with each other, and had an eye to the survival of the species and the planet, not just the fittest individuals.

Gaia considers the earth and its inhabitants to be a self-regulating system, the goal of which is 'the regulation of surface conditions so as to be as favourable as possible for contemporary life'.[7] Gaians are considered by some scientists as 'new age' fantasists, but Flannery argues that the theory is based on hard science.

Darwinism and its Medean outlook may provide solace to those unwilling to investigate the colonial past and its decimation of indigenous populations across the globe, but the future of the world and its creatures deserves our most coherent thought and judgement. To wonder about the trajectory of modern civilisations is not to sneer at private enterprise or scientific enquiry, but to wish those energies were directed in such a way that they do not destroy the planet.

This is not a bleeding-heart confection or adoration of the noble savage. It is prudent economic management and worshipful respect for the earth itself, the creation of

God or Bunjil or Buddha, it matters little which. Human survival on a healthy planet is not a soft liberal pipe dream; it is sound global management, and the deepest of religious impulses.

No peoples of the world in any era of their history wanted oppression, discomfort, or inconsequence. The desire for food, shelter, and purpose are universal; therefore, systems that provided citizens with as much of those three physical and psychological necessities must be considered successful and, furthermore, the survival of such systems over time can only have been wrought by the will of the people.

Anyone reading this book and the books that form part of its research will wonder why the trajectory of development in Aboriginal Australia did not lead to a full-blown scientific and agricultural advance. Rupert Gerritsen wonders if the innovation of any culture is in proportion to its size. This may explain the Australia that the explorers saw, but perhaps there is a philosophical, as well as evidentiary, reason for Aboriginal civilisation.

Maybe the destiny of mankind is still in flux, and the present iteration, of which we are so rightly in awe, has within its genius some dangerous flaws. The drive toward excellence, fuelled by the system of private enterprise, has an embedded need for exponential population growth, and, as we've experienced in the last few decades, this system seems incapable of protecting key resources such as air quality, fertile soils, and clean water.

It's not the difference between capitalism and communism; it's the difference between capitalism and Aboriginalism. Capitalism provides a platform for decisions among fellow capitalists, but shudders under the load of persuading communities over vast areas of the country. If that were not so, we would not have reached such an impasse with our management of the Murray-Darling basin; we would never consider leaving a state in our federation without drinking water; we would not have laws that allow coal-seam gas miners to ruin a farmer's land and threaten the very groundwater of the continent.

We keep telling ourselves that we are the lucky Australians, and we are right; but, as Donald Horne told us all those years ago, we are spending borrowed capital. We are still lucky — lucky to attend a music or sporting function with a crowd of 100,000 people, all of whom return home safely that night. We are still lucky to have sufficient food and quiet beaches, and we are lucky to have a working democracy.

To acknowledge the history of the country, and the social, agricultural, and philosophical achievements of Aboriginal and Torres Strait Islander people, does not put the economy at risk.

Restoring Aboriginal pride in the past and allowing that past to inform the future will remove the yoke of despair from Aboriginal people. Despair is reinforced every day an Aboriginal person has to argue for her pride in the past, or for his determination to honour the achievements of the ancestors. Ensuring that Aboriginal life and history are not wiped from the map because they interrupt the view from Parliament House will have a convulsive effect on the country's prospects.

Encouraging full participation of Aboriginal people is not a simple task of handing out fluorescent vests to work in a billionaire's mine, but requires a conversation with Aboriginal people about the future of the country. The opportunity to be involved in that future will release Aboriginal people from some of the shackles of colonialism. The country will still be colonised, but the dispossessed will be included, not just in the vote or constitution, but in the general Australian psyche. We will approach the idea of One Nation not by exclusion, but by an inclusion that rarely gets mentioned: Aboriginal participation.

More importantly, however, it will have intellectual and moral benefits, freeing us from the mental gymnastics we currently perform to rationalise colonialism and dispossession.

It seems improbable that a country can continue to hide from the actuality of its history in order to validate the fact that having said sorry, we refuse to say thanks. Should we ever decide to say thanks, the next step on a moral nation's agenda is to ensure that every Australian acknowledges the history and insists that, as we are all Australians, we should have the opportunity to share the education, health, and employment of that country on equal terms. Many will say that equality is insufficient to account for the loss of the land, but in our current predicament it is not a bad place to start.

The start of that journey is to allow the knowledge that Aboriginals did build houses, did cultivate and irrigate crops, did sew clothes, and were not hapless wanderers across the soil, mere hunter-gatherers. Aboriginals were intervening in the productivity of the country, and what they learnt during that process over many thousands of years will be useful to us today. To deny Aboriginal agricultural and spiritual achievement is the single greatest impediment to inter-cultural understanding and, perhaps, to Australian moral and economic prosperity.

Acknowledgements

I am deeply indebted to Rupert Gerritsen for his book *Australia and the Origins of Agriculture*. I found my way to that book via a Google search for Australian Aboriginal grain crops. It was listed number 35 in the search results, and all previous listings referred to contemporary wheat crops, an indication of the poverty of analysis devoted to this topic.

Later in my search, I came across an essay by Bill Gammage in which he talks about Aboriginal gardening and farming. Gammage's most recent book, *The Biggest Estate on Earth*, investigates, in exhaustive detail, the reports of early explorers and settlers, many of whom talked about the 'gentleman's estate' they had chanced upon. Not a wilderness, not a land peopled by wanderers, but a managed landscape created by the enormous labour of a people intent on creating the best possible conditions for food production.

I owe a great debt to my editor, Margaret Whiskin, for her discrimination and encouragement, and for allowing this book to live.

I'm grateful to the ancestors for their ingenious protection of the land. Where else on earth was there a civilisation that lasted more than 80,000 years and depended on both agriculture ... and peace?

Many other people provided information and advice, including Gordon Briscoe, John Clarke, Neville Oddie, Lyn Harwood, Susan Pascoe, Jack Pascoe, Koorie Heritage Trust, Victorian Aboriginal Corporation for Languages, Brewarrina Cultural Centre, Brad Steadman, Uncle Max Harrison, Sue Wesson, Lynnette Solomon-Dent, Vicki Couzens, Fran Moore, Ted Lefroy, Herb Patten, Michael Perry (cartographic sleuth), Reg Abrahams, Elizabeth Williams, Pauline Whyman, Mike Merrony, Richard (Cooma) Swain, Ted Donelan, Christina Eira, Stephen Morey, Max Allen, Ian Chivers, Charlotte Finch, Veronica Frail, Harry Allen, Paula Martin, The Artefact, Sue Norman, Betty Cruse, Liddy Stewart, Maria Brandl, Annette Peisley, Peter Gardiner, Ray Norris, John Morieson, Michael Walsh, Jason Stewart, Nick Connaughton (superphosphate), St Kilda Indigenous Nursery Co-operative (SKINC), Russell Mullet, Beth Gott, Barrie Pittock, Kevin Lowe, Yutaka Kobayshi, Elaine van Kempen, Eric Rolls, Peter

Gebhardt (cartographic sleuth), John Campbell (miller), Barbara Hart, Ben Shewry, Michael Westaway, Liz Warning, Peter Wlodarcszyk, Red Beard Bakery, Giorgio di Maria, James Hird, the ladies of Lake Mungo, and many others.

Picture credits

Front matter

p. i: Photograph by Barnaby Norris. From *Emu Dreaming: an introduction to Australian Aboriginal astronomy*, Ray and Cilla Norris, Emu Dreaming, 2009, p. 5

Chapter 1

p. 18: Illustration by J.H. Wedge, 'J.H.W. Native women getting tam bourn roots 27 August 1835'. From the *Todd Journal Andrew (alias William) Todd John Batman's recorder and his Indented Head journal 1985*. Latrobe section, State Library of Victoria, p. 70

p. 19: Photographs by Vicky Shukuroglou

p. 27: Photograph by Beth Gott

p. 27: Illustration by John Conran, University of Adelaide

p. 28: 'Perennial Grain Crops for High Water Use — the case for Microlaena stipoides.' C.L. Davies, D.L. Waugh and E.C. Lefroy, RIRDC publication number 05/024. Adapted from the *Tindale map, Aboriginal Tribes of Australia: their terrain, environmental controls, distribution, limits and proper names*, Australian National University Press, Canberra, 1974

p. 37: Photographs by Jonathon Jones

p. 48: Photograph by Lyn Harwood

p. 65: Photograph by Lyn Harwood

Chapter 2

p. 72: Ref 85/1286–722. Tyrrell Collection, Powerhouse Museum, Sydney
p. 77: Ref 85/1285–1135. Photograph by Henry King. Tyrrell Collection, Powerhouse Museum, Sydney
p. 88: Photograph by D.F. Thomson. Courtesy of the Thomson family and Museum Victoria
p. 90: Photograph by Connah and Jones, University of New England. Reproduced in Memmott, P., *Gunyah, Goondie and Wurley: the Aboriginal architecture of Australia*, UQP: 2007, p. 69
p. 95: Photograph by Stephen Mitchell

Chapter 3

p. 122: Photograph from Queensland Museum. Reproduced in Memmott, P.; *Gunyah, Goondie and Wurley: the Aboriginal architecture of Australia*, UQP: 2007, p. 19
p. 124: Photograph by Lyn Harwood
p. 127: From Smyth Papers, La Trobe Library, Melbourne. Reproduced in Memmott, P., *Gunyah, Goondie and Wurley: the Aboriginal architecture of Australia*, UQP: 2007, p. 197
p. 128: From Smyth Papers, La Trobe Library, Melbourne. Reproduced in Memmott, P., *Gunyah, Goondie and Wurley: the Aboriginal architecture of Australia*, UQP: 2007, p. 197
p. 135: Photograph by Lyn Harwood
p. 137: Taken from *Science of man and Journal of the Royal Anthropological Society of Australasia*, Vol. 2, 1899, p. 65. Reproduced in Young, M., *Aboriginal People of the Monaro* (2000) 2nd ed. (2003), NSW National Parks and Wildlife Service, p. 314
p. 139: Photographs by Jane Pye
p. 141: Image courtesy of the State Library of South Australia. SLSA: Special Collection 994T M682 – Plate 20 – *Burying-ground of Milmeridien*
p. 143 (top): Image courtesy of the State Library of South Australia.

SLSA: Special Collection 994T M682 – Plate 16 – *Tombs of a tribe*
p. 143 (bottom): Haddon Library of Archaeology and Anthropology, University of Cambridge

Chapter 4

p. 153: Haddon Library of Archaeology and Anthropology, University of Cambridge

Chapter 5

p. 170 (top): Photograph by Lyn Harwood
p. 170 (bottom): Photograph by Helen Stagoll

Chapter 7

p. 214 (left and right): Photographs by Lyn Harwood
p. 215 (left): Photograph by Jeanette Hope
p. 215 (right): Photograph by Lyn Harwood

Chapter 8

p. 220: Public domain
p. 222: *Angus McMillan with two Aboriginal Friends*, photograph held by the Latrobe Collection, State Library of Victoria

Bibliography

Aboriginal Affairs Victoria in conjunction with the Kerrup Jmara Elders Aboriginal Corporation, 'Lake Condah: heritage management plan & strategy', Aboriginal Affairs Victoria, Melbourne, 1993

Albrecht, Rev. F.W., *The Natural Food Supply of the Australian Aborigines*, Aborigines' Friends Association, Adelaide, 1884

Allen, H., 'Where the Crow Flies Backwards: man and land in the Darling Basin', unpublished thesis, Research School of Pacific Studies, ANU, Canberra, 1972

——, 'The Bagundji of the Darling Basin: cereal gatherers in an uncertain environment', *World Archaeology*, vol. 5, 1974, pp. 309–22

——, *Australia: William Blandowski's illustrated encyclopaedia of Aboriginal Australia*, Aboriginal Studies Press, Canberra, 2010

Altman, J., H. Bek, and L. Roach, 'Native Title and Indigenous Utilisation of Wildlife: policy perspectives', Centre for Aboriginal Economic Policy Research, ANU College of Arts & Social Sciences, Canberra, discussion paper 95/1995

Anderson, S., *Pelletier: the forgotten castaway of Cape York*, Melbourne Books, Melbourne, 2009

Andrews, A.E.J. (Ed.), *Stapylton with Major Mitchell's Australia Felix Expedition*, Blubber Head Press, Hobart, 1986

Archer, M., 'Confronting Crises in Conservation' in D. Lunney and C.

Kickman (Eds), *A Zoological Revolution: using native fauna to assist in its own survival*, Royal Zoological Society of New South Wales and the Australian Museum, 2002, pp. 12–52

Arkley, L., *The Hated Protector*, Orbit Press, Melbourne, 2000

Ashwin, A.C., *From Australia to Port Darwin with Sheep and Horses in 1871*, Royal Geographic Society of Australasia (SA), 1932

Barber, M. and S. Jackson, *Indigenous Water Values and Water Management on the Upper Roper River Northern Territory: history and implications for contemporary water planning*, National Water Commission, 2012

Barlow, A., *The Brothers Barmbarmbult and the Mopoke*, Macmillan, Melbourne, 1991

Barta, T., 'Mr Darwin's Shooters: on natural selection and the naturalizing of genocide', *Patterns of Prejudice*, vol. 39, no. 2, 2005, pp. 116–37

——, 'They Appear Actually to Vanish from the Face of the Earth: Aborigines and the European project in Australia Felix', *Journal of Genocide Research*, vol. 10, issue 4, 2008a, pp. 519–39

——, 'Sorry, and Not Sorry, in Australia: how the apology to the stolen generations buried a history of genocide', *Journal of Genocide Research*, vol. 10, issue 2, 2008b, pp. 201–14

——, 'Decent Disposal: Australian historians and the recovery of genocide' in D. Stone (Ed.), *The Historiography of Genocide*, Palgrave, Melbourne, 2008c

Basedow, H., *Knights of the Boomerang*, Hersperian Press, Carlisle, 2004

Batman, J., 'The Settlement of John Batman on the Port Phillip', from his own journal, George Slater, 1856

Beale, E., *Sturt: the chipped idol*, Sydney University Press, Sydney, 1979

The Bega Valley Shire, Bega Valley Shire Council, 1995

Bednarik, R.G., 'The Origins of Human Modernity', *Humanities*, vol. 1, 2012

Bellanta, M., 'Fabulating the Australian Desert: Australia's lost race romances, 1890–1908', *Philament*, no. 3, April 2004

Peter Bellwood, *First Migrants: ancient migration in global perspective*, John Wiley & Sons, London, 2013

Bennett, M., 'The Economics of Fishing: sustainable living in colonial New South Wales', *Aboriginal History*, vol. 31, 2007, pp. 85–102

Berndt, R. and C. Berndt, *The World of the First Australians: Aboriginal traditional life: past and present*, Aboriginal Studies Press, Canberra, 1999

Beveridge, P., various published and unpublished manuscripts, including *Courtenie and Kurwie* (Native Companion and Emu), supplied as copies by Victorian Aboriginal Corporation for Languages from library and press collections, Box 140/3, library stamped 1911

——, *The Aborigines of Victoria and the Riverina*, Lowden Publishing, Donvale, 2008 (re-issue of ML Hutchinson publication of 1889)

Bird, C. and R.E. Webb (Eds), *Fire and Hearth: forty years on: essays in honour of Sylvia Hallam*, records of the Western Australian Museum, supp. 79, Western Australian Museum, 2011

Bird-Rose, D., *Nourishing Terrains*, Australian Heritage Commission, Canberra, 1996

——, 'Exploring and Aboriginal Land Ethic', *Meanjin*, vol. 47, no. 3, 1998

Blay, J., 'Bega Valley Region Old Path Ways and Trails Mapping Project', Bega Valley Regional Aboriginal Heritage Study, 2005

——, 'The Great Australian Paradox', Eden Local Aboriginal Land Council, 2012

——, *On Track*, New South Books, Sydney, 2015

Briscoe, G.N., *Racial Folly: a twentieth-century Aboriginal family*, ANU E Press and Aboriginal History Incorporated, ANU, Canberra, 2010

Brock, D.G., *To the Desert with Sturt: a diary of the 1844 expedition,* Royal Geographical Society of Australasia, South Australian Branch, 1975

Brockwell, J., C.M. Evans, M. Bowman, and A. McInnes, 'Distribution, Frequency of Occurrence and Symbiotic Properties of the Australian Native Legume Trigonella Suavissima Lindl. and Its Associated Root-Nodule Bacteria', *The Rangeland Journal*, vol. 32, no. 4, pp. 395–406, 26 November 2010

Brodribb, W.A., *Recollections of an Australian Squatter, 1835–1883*, John Woods and Co., Sydney, 1883

Broome, R., 'The Great Australian Transformation', *Agora*, vol. 48, no. 4, 2013

Bulmer, J., *John Bulmer's Recollections of Aboriginal Life*, A. Campbell (Ed.), Museum Victoria, Melbourne, 2007

Bunjilaka Museum, Museum Victoria, Exhibit note, 2009

Butler, B., 'A Snapshot of My Life', Facebook, 2012

Candelo Historical Committee, *Candelo Recollects*, 1984

Cane, S., 'Australian Aboriginal Seed Grinding and Its Archaeological

Record: a case study from the Western desert' in D. Harris and G. Hillman (Eds) *Foraging and Farming: the evolution of plant exploitation*, Unwin Hyman, London, 1989, pp. 99–119

Cathcart, M., *The Water Dreamers*, Text Publishing, Melbourne, 2009

Chalmers, D., *Eight Moons to Midnight*, unpublished manuscript, 2012

Chivers, I., *Native Grasses*, Fourth Edition, Native Seeds, 2007

——, 'Splendour in the Grass: new approaches to cereal production', *The Conversation*, July 2012, http://theconversation.com/splendour-in-the-grass-new-approaches-to-cereal-production-8301

Chivers, I., R. Warrick, J. Bowman, and C. Evans, *Native Grasses Make New Products*, RIRDC, Canberra, June 2015

Christie, M.J., 'Aboriginal Science for an Ecologically Sustainable Future', *Australian Teachers' Journal*, March 1991

Clark, C.M.H., *Select Documents in Australian History*, Angus and Robertson, Sydney, 1965

Clark, I.D., 'The Journals of George Augustus Robinson, Vol. 2, Oct 1840 to August 1841', *Heritage Matters*, 1998

Clark, I.D. and T. Heydon, *Dictionary of Aboriginal Placenames of Victoria*, Victorian Aboriginal Corporation for Languages, Melbourne, 2002

Clarke, P., *Where the Ancestors Walked*, Allen and Unwin, Melbourne, 2003

——, *Aboriginal People and Their Plants*, Rosenberg, Dural, 2007

——, *Aboriginal Plant Collectors*, Rosenberg, Dural, 2008

Cleland, J.B. and T.H. Johnston, 'Notes on Native Names and Uses of Plants in the Musgrave Ranges Region', *Oceania*, vol. 8, 1936, pp. 208–215, 328–342

Cleland, J.B. and T.H. Johnston, 'Aboriginal Names and Uses of Plants in the Northern Flinders Ranges' in *Transactions of the Royal Society of South Australia*, vol. 63, 1939a, pp. 172–79

Cleland, J.B. and T.H. Johnston, 'Aboriginal Names and Uses of Plants at the Granites, Central Australia' in *Transactions of the Royal Society of South Australia*, vol. 63, 1939b, pp. 22–6,

Cobley, J., *Sydney Cove 1788*, Hodder and Stoughton, London, 1962

Connor, J., *The Australian Frontier Wars*, UNSW, Sydney, 2002

Cooper, D., *ABC Science News*, Flinders Rangers Rock Shelter, ABC, 4 November 2016

Cooper, W., 'The Changing Dietary Habits of 19th Century Australian

Explorers', *Australian Geographer*, vol. 28, 1997
Craw, C., 'Tasting Territory', *The Australian-Pacific Journal of Region Food Studies*, no. 2, 2012a
——, 'Gustatory Redemption?: colonial appetites, historical tales and the contemporary consumption of Australian native foods', *International Journal of Critical Indigenous Studies*, vol. 5, no. 2, 2012b
Crawford, I.M., 'Traditional Aboriginal Plant Resources', Australian Museum supp. no. 15, 1982
Cruse, B., L. Stewart and S. Norman, *Mutton Fish*, Aboriginal Studies Press, Canberra, 2005
Cundy, B.J., 'The Secondary Use and Reduction of Cylindro-Conical Stone Artifacts', *NT Museum of Arts and Sciences*, vol. 2, no. 1, 1985, pp. 115–27
D'Arcy, P., *The Emu in the Sky*, Natural Sciences and Technology Centre, Canberra, 1991
Davey, M., *Brown Judy*, Penfolk, Melbourne, 2010
Davies, C.L., D.L. Waugh, and E.C. Lefroy, *Perennial Grain Crops for High Water Use*, Rural Industries Research and Development Corporation, Canberra, 2005a
——, 'Variation in Seed Yield and Its Components in the Australian Native Grass, *Microlaena stipoides*', *Australian Journal of Agricultural Research*, no. 56, 2005b
Davis, J., *Tracks of McKinlay and Party Across Australia*, Samson Low, London, 1863
Davis, M., 'Sealing and Whaling in Twofold Bay', unpublished manuscript, 2004
Davis, W., *The Wayfinders: why ancient wisdom matters in the modern world*, Anansi Press, Toronto, 2009
Dargin, P., *The Aboriginal Fisheries of the Darling-Barwon Rivers*, Brewarrina Historical Society, 1976
Dawkins, R., *The God Delusion*, Transworld, Black Swan, London, 2006
Dawson, J., *Australian Aborigines*, George Robertson, Melbourne, 1881
Denham, T., M. Donohue, and S. Booth, 'Horticultural Experimentation in Northern Australia Reconsidered', *Antiquity*, no. 83, 2009
Ditchfield, C., *Salting in Australia*, unpublished manuscript, 2013
Dix, W.C. and M.E. Lofgren, 'Kurumi: possible Aboriginal incipient agriculture', *Records of West Australian Museum*, vol. 3, 1974

Duncan-Kemp, A., *Our Sandhill Country*, Angus and Robertson, Sydney, 1934
Edwards, W.H. (Ed.), *Traditional Aboriginal Society*, Macmillan, South Yarra, 1987
Egan, J., *Australian Geographic*, December 2012, pp. 50–61
Eriksen, R., *Ernest Giles, explorer and traveller, 1835–97*, Heinemann, Melbourne, 1978
Eyre, E., *Reports of an Expedition to King Georges Sound*, Sullivan's Cove Press, Adelaide, 1983
Favenc, E., *The Secret of the Australian Desert*, Blackie, London, 1896
Field, B. (Ed.), *Geographical Memoirs of NSW*, J. Murray, London, 1825
Fisk, E.K., *The Aboriginal Economy*, Allen & Unwin, Sydney, 1985
Flannery, T., *The Explorers*, Text Publishing, Melbourne, 1998
——, *Here on Earth*, Text Publishing, Melbourne, 2010
Flood, J., *The Moth Hunters*, AIATSIS, Canberra, 1980
——, *Archaeology of the Dreamtime*, Collins, Sydney, 1983
Forbes, S. and L. Liddle, 'Hidden Gardens: Australian Aboriginal people and country', *The Good Gardener*, Artifice Books on Architecture, London, 2015
Frankel, D., 'An Account of Aboriginal Use of the Yam Daisy', *The Artefact*, vol. 7, nos 1–2, 1982, pp. 43–5
Fullagar, R. and J. Field, *Antiquity*, vol. 71, no. 272, 1997
Gammage, B., 'Australia Under Aboriginal Management', Barry Andrews Memorial Lecture, ANU, 2002
——, 'Gardens without Fences', *Australian Humanities Review*, Issue 36, July 2005
——, *Galahs*, unpublished manuscript, 2008
——, *Australian Historical Studies*, April 2011, vol. 31, 'The History of Gardens and Designed Landscapes.'
——, *The Biggest Estate on Earth*, Allen & Unwin, Sydney, 2011
Genoa Town Committee, *Border Tales*, Genoa Town Committee, 2000
Gerritsen, R., *And Their Ghosts May Be Heard*, Fremantle Arts Centre Press, Fremantle, 1994
——, 'Nhanda Villages of the Victoria District of WA', Intellectual Property Publications, Canberra, 2002
——, *Australia and the Origins of Agriculture*, Archaeopress, London, BAR series, 2008

——, *Beyond the Frontier: explorations in ethnohistory*, Batavia Online Publishing, Canberra, 2011

Gibbs, M., *An Aboriginal Fish Trap on the Swan Coastal Plain*, Records of W.A. Museum, supp. no. 79

Giles, E., *Australia Twice Traversed*, Gutenberg, 1872–1876

Gill, I., *All That We Say is Ours*, Douglas and Macintyre, Vancouver, 2010

Gillespie, W.R., 'The Northern Territory Intervention and the Mining Industry', unpublished manuscript, 2009a

——, 'Infamy of the Intervention', unpublished manuscript, 2009b

Gilmore, M., 'Fish Traps and Fish Balks', *Sydney Morning Herald*, 8 November 1933

——, *Old Days, Old Ways*, Angus and Robertson, Sydney, 1934

Goad, P. and J. Willis, *The Encyclopedia of Australian Architecture*, Cambridge University Press, Port Melbourne, 2011

Gorecki, P. and M. Grant, 'Grinding Patches from the Croydon Region, Gulf of Carpentaria', *Archaeology in the North*, North Australia Research Unit, ANU, Canberra, 1994

Gott, B., 'Plant Resources of Mallacoota Area', Series: Rep. 82/31, 1982

——, 'Murnong—Microseris scapigera: a study of a staple food of Victorian Aborigines', *Australian Aboriginal Studies*, vol. 2, 1983, pp. 2–18

——, 'Murnong: a Victorian staple food', *Archaeology*, ANZAAS, 1986

——, 'Ecology of Root Use by the Aborigines of Southern Australia', *Archaeology in Oceania, no.* 19, 1991

——, 'Cumbungi Typha: a staple Aboriginal food in Southern Australia', *Australian Aboriginal Studies*, no. 1, 1999

——, 'Fire-Making in Tasmania', *Current Anthropology*, vol. 43, no. 4, 2002

——, 'Aboriginal Fire Management in S.E. Australia: aims and frequency', *Journal of Biogeography*, no. 32, 2005

Gott, B. and J. Conran, *Victorian Koorie Plants*, Yangennanock Women's Group, Hamilton, 1991

Gott, B. and N. Zola, *Koorie Plants, Koorie People: traditional Aboriginal food, fibre, and healing plants of Victoria*, Koorie Heritage Trust, Melbourne, 1992

Gould, R., *Yiwara*, Scribners, New York, 1969

Goyder, G.W., 'Northern Exploration', Parliamentary Papers, 1857-8SA

No. 72/1857, pp.1–4, in Gerritsen, R., *Australia and the Origins of Agriculture*, Archaeopress, London, BAR series, 2008

Graham, C., 'Telling Whites What They Want to Hear', *Overland*, Issue 200, 2010

Gray, S., *The Protectors*, Allen & Unwin, Sydney, 2011

Gregory, A.C., *Journals of Australian Explorations*, James Beal, Gov. Printer, Brisbane (facsimile ed.), 1884

——, 'Memorandum on the Aborigines of Australia', *Journal of Anthropology of Great Britain and Ireland*, vol. XV and XVI, 1887

Grey, G., from Outbackvoices.com, 2009

Grigg, G., P. Hale, and D. Lunney, 'Kangaroo Harvesting in the Context of Ecologically Sustainable Development and Biodiversity Conservation', *Conservation Biology*, University of Queensland Press, 1995

The Guardian, 5 September 2016

Hagan, S., *The N Word*, Magabala Books, Broome, 2005

Hallam, S., *Fire and Hearth: a study of Aboriginal usage*, AIAS, Canberra, 1995

Harney, B., *North of 23°*, Australasian Publishing Company, Sydney, 1946

Hart, C.W.M. and A.R. Pilling, *The Tiwi of Northern Australia*, Holt Rinehart and Winston, New York, 1979

Hawley, J., 'Art Masters', *Good Weekend*, 4 September 2010

Heaney, S. (trans.), *Beowulf*, Faber, London, 1999

Henderson, B. et al., 'More Than Words', *Language, Documentation and Conservation* vol. 8, 2014

Henry, R., et al., *Australian Orysa: utility and conservation*, Springer Science, 2009

Henson, C.F., *Telling Absence: Aboriginal social history and the national museum of Australia*, ANU, 2009

Hercus, L.A., F. Hodges and J.H. Simpson, *The Land is a Map*, ANU, Canberra, 2002

Hiddins, L., *Bush Tucker Man*, ABC Books, Melbourne, 2001

Hill, B., *Broken Song*, Knopf, Sydney, 2002

Hinkson and Beckett (Eds), *An Appreciation of Difference*, WEH Stanner and Aboriginal Studies Press, Canberra, 2008

Hoare, M., *The Half Mad Bureaucrat*, Records of Australian Academy of Science, Canberra, 1973

Hoffman, P.T., *Why Did Europe Conquer the World*, Princeton University Press, 2015

Hope, J. and G. Vines, *Brewarrina Aboriginal Fisheries Conservation Plan*, Brewarrina Aboriginal Cultural Museum, 1994

Howe, K.R. (Ed.), *Mallacoota Reflections*, Mallacoota and District Historical Society, 1990

Howitt, A.W., *Land, Labour and Gold*, vols 1 and 2, Sydney University Press, 1855

——, 'On Songs and Songmakers of Some Australian Tribes', *Journal of Anthropology of Great Britain and Ireland*, vol. XV, 1886

Hunter, J., *An Historical Journal of the Transactions at Port Jackson and Norfolk Island*, Stockdale, London, 1792

——, *An Historical Account of the Transactions at Port Jackson and Norfolk Island*, Libraries Board of South Australia, 1968

Hutchings, R. and M. La Salle, 'Teaching Anti-colonial Archaeology', *Archaeologies* vol. 10, no. 1, 2014.

Hynes, R.A. and A.K. Chase, 'Plants, Sites and Domiculture: Aboriginal influence upon plant communities in Cape York Peninsula', *Oceania*, vol. 17, 1982

Israeli, R., *Poison: modern manifestations of a blood libel*, Lexington Books, New York, 2009

Johnston, A., and M. Rolls, *Reading Robinson*, Quintus, Hobart, 2008

Jones, B.T., 'Embracing the Enemy', *Australian Geographic*, no. 116, 2013

Keen, I., *Aboriginal Economy and Society*, OUP, Melbourne, 2004

——, 'The Brief Reach of History', *Oceania*, vol. 76, no. 2, 2006

——, *Variation in Indigenous Economy and Society at the Threshold of Colonisation*, AIATSIS, Canberra, 2008

Kenyon, A.S., 'Stone Structures of the Australian Aborigines', *The Victorian Naturalist*, no. 47, Kerrup Jmara Elders and AAV, Lake Condah: Heritage Management Plan & Strategy, 1993

Kershaw, A.P., 'A Quartenary History of N.E. Queensland from Pollen Analysis', *Quartenary Australasia*, vol. 12, no. 2, 1994

Kimber, R.G., *Resource Use and Management in Central Australia*, Australian Aboriginal Studies, Canberra, 1984

— 'Beginnings of Farming', *Mankind*, vol. 10, no. 3, June 1976

Kirby, J., *Old Times in the Bush of Australia: trials and experiences of early bush life in Victoria: during the forties*, G Robertson and Company, Victoria, 1897

Koch, H. and L. Hercus (Eds), *Aboriginal Placenames*, ANU E Press, Canberra, 2008

Kohen, J., 'The Impact of Fire: an historical perspective', paper presented at SGAP Biennial Seminar, 1993

—— (Ed.), *Aboriginal Environment Impacts*, UNSW Press, Sydney, 1995

Kondo, T., M.D. Crisp, C. Linde, D.M.J.S. Bowman et al., 'Not An Ancient Relic: the endemic *Livistona* palms of arid central Australia could have been introduced by humans', *Proceedings of the Royal Society*, 7 July 2012, vol. 279, no. 1738, pp. 2652–61

Koori Mail, 18 May 2016

Latz, P.K., *Pocket Bushtucker*, IAD Press, Alice Springs, 1999

— *Bushfires and Bushtucker*, IAD Press, Alice Springs, 1995

Laurie, V., *The Monthly*, Melbourne, June 2011

Lawlor, R., *Voices of the First Day*, Inner Traditions, New York, 1991

Le Griffon, H., *Campfires at the Cross*, Australian Scholarly Publishing, Melbourne, 2006

Lindsay, D., *Explorations in the Northern Territory of South Australia*, Royal Geographical Society of Australasia (South Australia), 1890

Lingard, J., *A Narrative of a Journey to and from NSW*, J. Taylor and co, Chapel-en-le-Frith, 1846

Long, A., *Aboriginal Scarred Trees in New South Wales: a field manual*, Department of Environment and Conservation NSW, 2005

Lourandos, H. and A. Ross, 'The Great Intensification Debate: its history and place in Australian archaeology', *Archaeology*, no. 39, 1994

——, *Continent of Hunter Gatherers*, Cambridge University Press, Cambridge, 1997

Lowe, P. and J. Pike, *You Call it Desert — We Used to Live There*, Magabala Books, Broome, 1980

Macinnis, P., *Australia: pioneers, heroes, and fools*, Pier Nine, Murdoch Books, Sydney, 2007

Maggiore, P.M.A., 'Utilisation of Some Australian Seeds in Edible Food Products' in *The Food Potential of Seeds from Australian Native Plants* (proceedings of a colloquium held at Deakin University on 7 March 1984), G.P. Jones (Ed.), Deakin University Press, Melbourne, 1985 pp. 59–74

Mallacoota Historical Society, *Mallacoota Reflections*, Mallacoota Historical Society, 1990
Manton, B., 'A National Disaster We Choose to Ignore', Drum, ABC, 2011
Maslen, G., 'Cutting Edge', *The Age*, 10 February 2010
Mate Mate, R., *Barndana*, unpublished manuscript
Mate Mate, R., *A Brief Insight into the Wurunjeri Tribe: the uncompleted chapter*, unpublished manuscript, Attorney-General's Department, 1989a
——, 'A Tribute to Winnie Narrandjerri Quagliotti', Wurundjeri Tribal Land and Cultural Heritage Council Inc, 1989b
Mathews, R.H., 'Aboriginal Fisheries at Brewarrina', *Journal of the Royal Society of NSW*, 1903
McCarthy, F., 'The Grooved Conical Stones of New South Wales', *Mankind*, vol. 2, no. 6, 1939
McConvell, P., *A Short Ride in a Time Machine*, MUP, Melbourne, 2004
McConvell, P. and N. Evans, (Eds), *Archaeology and Linguistics*, OUP, Melbourne, 1997
McConvell, P. and M. Laughren, 'Millers, Mullers and Seed Grinding', in H. Anderson (Ed.), *Language Contacts in Prehistory: studies in stratigraphy*, John Benjamins Publishing Company, Amsterdam, 2003
McKinlay, J., *McKinlay's Journal of Exploration in the Interior of Australia (Burke Relief Expedition)*, F.F. Bailliere, Melbourne, 1862
McMillan, A., *An Intruder's Guide to Arnhem Land*, Duffy and Snellgrove, Sydney, 2001
McNiven, I., and L. Russell, *Appropriated Pasts*, Altamira Press, Oxford, 2005
Memmott, P., *Gunyah, Goondie and Wurley: the Aboriginal Architecture of Australia*, UQP, Brisbane, 2007
Menzies, G., *1421: the year China discovered the world*, Bantam, Transworld, London, 2002
Mitchell, T.L., *Three Expeditions into the Interior of Eastern Australia*, vols 1 and 2, T and W Boone, London, 1839
——, *Journal of an Expedition into the Interior of Tropical Australia*, Greenwood Press, New York, 1969
Mollison, B., 'A Synopsis of Data on Tasmanian Aboriginal People', unpublished paper, Psychology Department, University of Tasmania, Hobart, 1974
Morcom, L. and M. Westbrooke, 'The Pre-Settlement Vegetation of the

Western and Central Wimmera Plains', *Australian Geographical Studies*, November 1998
Morgan, P., *Foothill Farmers*, Ngarak Press, Ensay, Victoria, 2010
Morieson, J., 'Aboriginal Stone Arrangement in Victoria', unpublished paper, Australian Centre, University of Melbourne, 1994
——, 'Rock Art and Indigenous Astronomies', unpublished paper, 3rd AURA Conference, Alice Springs, 2000
——, *Solar Based Lithic Design*, World Archaeological Congress, Washington, 2003
——, *Munungabumbum of the Dja Dja Wurrung: the mentor of Mindi*, unpublished manuscript, 2010
Muir, C., *Writing the Toad*, unpublished manuscript, 2011
Mulvaney, K., *Burrup and Beyond*, Ken Mulvaney, Perth, 2013
Mulvaney, K. et al. (Eds), *My Dear Spencer*, Hyland House, Melbourne, 2000
Museum Victoria, *Bunjilaka*, Museum Victoria, 2000
Nakata, N., *The Cultural Interface*, PhD thesis, James Cook University, 2007
Niewojt, L., 'Gadabanud Society in the Otway Ranges', *Aboriginal History*, no. 33, 2009
Norris, R. and C. Norris, *Emu Dreaming*, Emu Dreaming, Sydney, 2009
O'Connell J., P. Latz, and P. Barnett 'Traditional and Modern Plant Use Among the Alyawara of Central Australia', *Economic Botany*, no. 37, 1983, pp. 80–109
O'Connor, N. and K. Jones, *A Journey Through Time*, self-published family history, 2003
O'Mara, P. (Ed.), *Medical Journal of Australia*, vol. 192, no. 10, Australian Medical Publishing, Sydney, May 2010
Organ, M.K. and C. Speechley, *Illawarra Aborigines — An Introductory History*, University of Wollongong, 1997
Pascoe, B., *Convincing Ground*, Aboriginal Studies Press, Canberra, 2007
Peisley, A., *Our Outback Larder*, unpublished manuscript, 2010a
——, *The Re-Discovery of Gippsland Explorers*, unpublished manuscript, 2010b
——, *A–Z Plants*, unpublished manuscripts, 2011
——, various unpublished research documents (Pascoe collection)

Pepper, P. and T. DeAraugo, 'The Kurnai of Gippsland', *Hyland*, vol. 1, 1985

Phillips, G., 'Life Was Not a Walkabout for Victoria's Aborigines', *The Age*, 13 March 2003

Platts, L., *Bygone Days of Cathcart*, Platts, Cathcart, 1989

Pope, A., *One Law for All?*, unpublished manuscript, 2010

Poulson, B., *Recherche Bay*, Southport Community Centre, 2004

Read, P. (Ed.), *Indigenous Biography and Autobiography*, ANU E Press and Aboriginal History Incorporated, Aboriginal History Monograph 17, 2008

Reynolds, H., *An Indelible Stain?: the question of genocide in Australia's history*, Viking, Ringwood, 2001

Rhea, R., 'Knowing Country, Knowing Food', *Artefact*, vol. 35, 2012

Rifkin, J., *Beyond Beef: the rise and fall of the cattle culture*, Penguin, New York, 1991

Robinson, G.A., *The Journals of George Augustus Robinson*, vol. 2, 31 October 1841, Heritage Matters, Melbourne, 1998

Rolls, E., *A Million Wild Acres*, Nelson, Melbourne, 1981

——, 'A Song of Water', *Island*, no. 102, Spring 2005

——, *Epic Rolls*, unpublished manuscript, 2009

—— (ed.), *An Anthology of Australian Fishing*, McPhee-Gribble, Ringwood, 1991

Rose, F., *The Traditional Mode of Production of the Australian Aborigines*, Angus and Robertson, Sydney, 1987

Rose, N. (ed.), *Fair Food*, UQP, Brisbane, 2015

Ross, P., Ngarrindjeri *Fish Traps of the Lower Murray Lakes and Northern Coorong Estuary*, PhD thesis, Flinders University, Adelaide, 2009

Roth, H.L., 'On the Origins of Agriculture', *Journal of Anthropology of Great Britain and Ireland*, vol. XV, 1886

Royal Botanical Gardens of Victoria, 2016, *Muellaria*, vol. 34, pp. 63–7

Russell, L. and I. McNiven, 'Monument Colonialism', *Journal of Material Culture*, vol. 3, no. 3, 1998

Russell-Smith, J. et al, 'Aboriginal Resource Utilization and Fire Management in Western Arnhem Land, Monsoonal Northern Australia, Notes for Prehistory, Lessons for the Future', *Human Ecology*, vol. 25, no. 2, 1997

Sacks, O., *The Island of the Colourblind and Cycad Island*, Vintage Books, Random House, New York, 1998

Samson, B., 'The Brief Reach of History and the Limitation of Recall', *Oceania*, no. 76, 2006
Scofield, C., *Bombala: Hub of Southern Monaro*, Shire of Bombala, 1990
Scott, G.F., *The Last Lemurian*, James Bowden, London, 1898
Sefton, C., 'Molluscs and Fish in the Rock Art of the Coast, Estuary and Hinterland of the Woronora Plateau of NSW', *Rock Art Research*, vol. 1, no. 2, 2011
Smith, B., '35,000 Year Old Axe Head', *The Age*, 12 November 2010
Smith, J. and P. Jennings, 'The Petroglyphs of Gundungurra Country', *Rock Art Research*, vol. 28, no. 2, 2011
Smith, L.T., *Decolonising Methodologies*, Zed Books, London, 2012
Smyth, D., 'Saltwater Country: Aboriginal and Torres Strait Islander Interest in Ocean Policy Development', Socio-cultural Policy Paper 36, Commonwealth of Australia, 1997
Smyth, R.B., *The Aborigines of Victoria and the Riverina*, John Ferres, Gvt. Printer, Melbourne, 1878
Snyder, T., *Bloodlands: Europe between Hitler and Stalin*, Vintage, London, 2011
Spencer, B. and F. Gillen, *The Native Tribes of Central Australia*, Dove, New York, 1899
Stanner, W.E.H., *White Man Got No Dreaming: essays, 1938–73*, ANU, Research School of Social Sciences, 1979
——, *After the Dreaming: black and white Australians — an anthropologist's view*, Australian Broadcasting Commission, Boyer lectures, Sydney, 1968
State Library of NSW, *Marinawi: Aboriginal odysseys*, State Library of NSW, 2010
Stuart, J.M., *Explorations in Australia: the journals of John McDouall Stuart, Saunders and Otley*, 1864
Sturt, C., *Two Expeditions into the Interior of Southern Australia*, vols I and II, Smith, Elder, London, 1833
——, *Narrative of an Expedition into Central Australia*, T & W Boone, 1849
Sullivan, M., S. Brockwell, and A. Webb, *Archaeology in the North*, proceedings of the 1993 Australian Archaeological Association Conference, 1994
Sullivan, P., 'Desert Knowledge', working paper no. 4, Indigenous Governance, Desert Knowledge CRC, 2007

Tatz, C., 'Genocide in Australia', *AIATSIS Research Discussion Papers*, no. 8, 2000

Thomas, W., unpublished transcription of his papers by the Victorian Aboriginal Corporation of Languages, 2013

Thomson, D., *Donald Thomson in Arnhem Land*, Miegunyah Press, Melbourne, 2003

Tindale, N., 'Adaptive Significance of Panara or Grass Seed Culture of Australia' in Wright, B., *Hunting and Gathering and Fishing*, AIATSIS, Canberra, 1978

Todd, A., *The Todd Journal 1835* (and publisher's notes), Library Council of Victoria, 1989

Tonkin, D. and C. Landon, *Jackson's Track*, Penguin, Melbourne, 1999

University of Western Australia, 'Recording a Visual History', *Uniview*, vol. 30, no. 1, 2011

Tales of Far South Coast, *Tales of Far South Coast Journal*, vols 1–4, Merimbula, 1982

Turnbull, D., *To Talk of Many Things*, IT University, Copenhagen, September 2015

Weatherhead, A., *Leaves from My Life*, Eden Museum, (facs.) 1998

Wakefield, N., 'Bushfire Frequency and Vegetational Change in SE Australian Forests', *Victorian Naturalist*, no. 87, 1970

Walsh, N., 'A Name for Murnong', *Royal Botanic Gardens*, no. 34, 2016, pp. 63–7

Walters, I., 'Some Observations on the Material Culture of Aboriginal Fishing in the Moreton Bay Area: implications for archaeology', *Queensland Archaeological Research*, vol. 2, University of Queensland, Brisbane, 1985

Wesson, S., *Gippsland Women's Oral History*, Gippsland and East Gippsland Aboriginal Co-op, 1997a

——, *Transcripts and Extracts Taken from Records for NE Victoria and Southern NSW*, unpublished manuscript, 1997b

——, 'An Historical Atlas of the Aborigines of Eastern Victoria and South-eastern NSW', *Monash Publications in Geography and Environmental Science*, no. 53, 2000

——, 'The Aborigines of Eastern Victoria and Far Southeastern New South Wales, 1830 to 1910: an historical geography', unpublished thesis, Monash University, Faculty of Arts, School of Geography and Environmental Science, 2003

Westminster Select Committee of Legislative Council, 'The Aborigines', Victorian Senate papers office, 3 February 1859

White, J.P., 'Agriculture: was Australia a bystander?', presented at The Fifth World Archaeological Congress Washington DC, 2003

White, P., 'Revisiting the Neolithic Problem in Australia', in Bird and Webb, *Fire, Hearth: forty years on,* Records of Western Australian Museum, supp. 79

Williams, E., 'Documentation and Investigation of an Aboriginal Village in South Western Victoria', *Aboriginal History*, vol. 8, 1984

——, 'Complex Hunter Gatherers', *Antiquity*, vol. 61, 1987

——, *The Archaeology of Lake Systems in the Middle Cooper Basin*, records of the South Australian Museum, 1998

Williams, J., *Clam Gardens: Aboriginal mariculture on Canada's west coast (Transmontanus)*, New Star Books, Vancouver, 2006

Willingham, R., 'Native Title Law Attacked', *The Age*, 16 September 2010

Woolmington, J., *Aborigines in Colonial Society*, Cassell Australia, North Melbourne, 1973

Wright, B., 'The Fish Traps of Brewarrina', Aboriginal Health Conference, NSW, September 1983

Wurm, P., L. Campbell, G.D. Batten, and S.M. Bellairs, 'Australian Native Rice: a new sustainable wild food enterprise', Research Institute for the Environment and Livelihoods, Rural Industries Research and Development Corporation, Research Project No. PRJ000347/ Publication No. 10/175, 2012

Young, M. (compiler), *The Aboriginal People of the Monaro*, second edition, Department of Environment and Conservation (NSW), 2005

Notes

Introduction

1 Barta, 2005, p. 124
2 Smith, L.T., pp. 20–4
3 Kirby, pp. 31–2
4 ibid., pp. 35–6
5 ibid., p. 36
6 Beveridge, 1889/2008, pp. 54, 103–6
7 Kirby, p. 79
8 ibid.
9 ibid., p. 109
10 Rolls, 1981, p. 84

Chapter 1: Agriculture

1 Gerritsen, 2008, pp. 39–41, 62
2 Gammage, 2011, p. 281
3 Mitchell, 1848/1969, p. 90
4 Mitchell, 1839, vol. 1, pp. 237–38
5 Mitchell, 1839, vol. 2, p. 194
6 Andrews, p. 146
7 Grey, pp. 6–7

8 Gerritsen, 2008, p. 33
9 Robinson, p. 326
10 Mitchell, 1839, (from Gott, 2005, p. 1204)
11 Hunter, 1793/1968
12 Arkley, p. 317
13 Batey quoted in Frankel, p. 44
14 Gott, 1991, p. 65
15 Frankel, pp. 43–4
16 Le Griffon, p. 51
17 ibid.
18 Gott, 2005, p. 1205
19 Gammage, 2011, p. 190
20 Mitchell, 1839, vol. 1, p. 90
21 Mitchell, 1848/1969, p. 274
22 Mitchell, 1839, vol. 1, p. 14
23 Rolls, 1981, p. 37
24 Howitt, W, 1855, p. 309
25 Kimber, 1984, p. 16
26 Gerritsen, 2008, p. 60
27 Rolls, 2009, Ch. 7, p. 7
28 ibid.
29 Duncan-Kemp, pp. 146–7
30 Robinson 1841/1998, vol. 2, p. 207
31 ibid.
32 Mitchell, 1839, vol. 1, p. 238
33 Sturt, 1849, p. 69
34 ibid., p. 71
35 Brock, p. 133
36 Sturt, 1849, p. 155
37 Mitchell, 1839, vol. 2, p. 65
38 Morcom and Westbrooke, p. 286
39 Kimber, 1984, p. 15
40 McKinlay, 1861, in Gerritsen, 2008, p. 50
41 Gerritsen, 2008, p. 43
42 Etheridge, 1894, in Gerritsen, 2008, p. 110

43 Gerritsen, 2008, pp. 42, 78
44 ibid., p. 83
45 ibid., p. 84
46 Gammage, 2008, p. 14
47 Chivers, 2012
48 ibid., p. 2
49 ibid., pp. 2, 4
50 Wurm et al., 2012, p. 1
51 Kimber, 1984, p. 19
52 Rolls, 2005, p. 15
53 Gorecki and Grant, pp. 235–6
54 Sullivan, Brockwell, and Webb, pp. 235–6
55 ibid., p. 16
56 Gerritsen, 2011, p. 25
57 Rolls, 2005, p. 15
58 ibid.
59 Sturt, 1849, p. 90
60 John Morieson, personal conversation and demonstration with the author, 2009
61 Dix and Lofgren, pp. 73–7
62 Barber and Jackson, pp. 18–50
63 Mitchell, 1839, vol. 2, p. 153
64 Gammage, 2011, p. 132
65 Archer, p. 20
66 Gerritsen, 2008, p. 50
67 Latz, 1999, p. 17
68 Ashwin, 1870–71 in Gammage, 2008, p. 5
69 Harney, p. 45
70 Gerritsen, 2008, p. 45
71 Denham et al., p. 637
72 Kirby, p. 28
73 ibid., p. 34
74 Mitchell, 1839, vol. 2, p. 61
75 Kirby, p. 28
76 Mitchell, 1839, vol. 2, p. 134

77 Courtenie and Kurwie, p. 1, in Beveridge, 1911
78 Beveridge, 'The Story of Coorongendoo Muckie of Balaarook', 1911, p. 3
79 Gerritsen, 2008, p. 60
80 Denham et al., p. 643
81 Tindale, pp. 345–9
82 ibid., p. 141
83 *Koori Mail*, 18 May 2016, p. 4
84 Davis, W., pp. 8–9
85 Cooper, D.
86 Kershaw, pp. 1–11
87 Ross, p. 29
88 Lourandos, 1997, p. 335
89 Lourandos, 1994, p. 60
90 Grigg, Ch. 26
91 ibid.
92 Latz, 1995, pp. 54–5

Chapter 2: Aquaculture

1 Beveridge, 1889/2008, p. 89
2 ibid.
3 Gibbs, p. 6
4 Wright, p. 3
5 Cruse and Norman, p. 17
6 Walters, p. 51
7 Stuart, p. 68
8 Mathews, pp. 146–56 and Dargin, p. 38
9 Wikipedia, 2013 and NSW Heritage Council, 15 April 2010
10 Hope and Vines, p. 67
11 Phillips
12 Batman
13 Morieson, 1994, p. 34
14 Thomas
15 Wesson, 2000, pp. 91–2
16 Gerritsen, 2008, p. 111

17 Williams, J., p. 18
18 ibid., p. 20
19 ibid., p. 118
20 ibid.
21 Dawson, p. 19
22 Sturt, 1849, p. 111
23 Mitchell, 1839, vol. 1, p. 336
24 Gilmore, 1933
25 Davis, J., p. 90
26 Memmott, p. 68
27 Melbourne Museum, 2009 [no ref?]
28 Cruse et al., p. 8
29 Beveridge, 1889/2008, p. 95
30 Smyth, D., p. 6

Chapter 3: Population and Housing

1 Sturt, 1849, p. 111
2 ibid.
3 ibid., p. 58
4 ibid., p. 108
5 ibid., p. 124
6 Gerritsen, 2011, p. 25
7 ibid., p. 29
8 Organ and Speechley, p. 6
9 Duncan-Kemp, 1934
10 Willingham
11 Mitchell, 1848/1969, p. 90
12 Sturt, 1833, vol. 1, p. 298; vol. 2, p. 140
13 Memmott, p. 223
14 ibid., p. 22
15 Gammage, 2011, p. 229
16 ibid.
17 ibid., p. 231
18 Mitchell, 1839, vol. 1, pp. 76–7
19 ibid., p. 240

20 ibid., vol. 2, p. 247
21 ibid., p. 351
22 ibid., vol. 1, pp. 156–8
23 ibid., p. 160
24 ibid., p. 90
25 Barta, 2008a, p. 520
26 Mitchell, 1839, vol. 2, pp. 96–7
27 Andrews, 1986, p. 77
28 Gerritsen, 2008, p. 164
29 Mitchell, vol. 2, p. 271
30 Pope, p. 12
31 Lindsay, p. 4
32 Goyder, p. 4
33 Stuart, pp. 42, 71
34 McMillan, p. 46
35 Howe, p. 10
36 Mitchell, 1839, vol. 1, p. 225
37 Gerritsen, 2008, p. 50
38 Williams, E., 1984, p. 174
39 Memmott, p. 166
40 ibid., pp. 170–8
41 ibid., p. 136
42 ibid., p. 74
43 Field, p. 54
44 Le Griffon, p. 291
45 *The Guardian*, 15 September 2016
46 Peisley, unpublished research documents (Pascoe collection)
47 Dawson, p. 10
48 Kenyon, pp. 71–5
49 McNiven and Russell, p. 113
50 Memmott, p. 185
51 Johnston and Rolls, p. 137
52 Mollison, pp. 7.2–7.3
53 Thomson, p. 217
54 Memmott, p. 223

55 ibid., p. 237
56 Peisley, unpublished research documents (Pascoe collection)
57 Mitchell, 1839, vol. 1, p. 321

Chapter 4: Storage and Preservation

1 Gerritsen, 2008, p. 55
2 ibid., p. 56
3 Ashwin, 1932, p. 64 (from Gerritsen, 2008, p. 57)
4 Howitt in Smyth, R.B., vol. 2, pp. 302–3
5 Gerritsen, 2008, pp. 56–7
6 Crawford, p. 8
7 Young, p. 246
8 Flood, 1980, p. 74
9 Peisley papers, Pascoe collection and Flood, 1980, p. 81
10 Gerritsen, 2008, p. 82
11 Gerritsen, 2011, p. 58
12 Gerritsen, 2008, pp. 56, 79
13 ibid., p. 81
14 Hutchings and La Salle, pp. 34–35
15 Egan, pp. 50–61
16 ibid.

Chapter 5: Fire

1 Mitchell, 1848, quoted by Gott, 2005, p. 204
2 Gammage, 2011, p. 338
3 ibid., p. 242
4 Wakefield, p. 138
5 Kohen, 1993, p. 4
6 ibid., p. 5
7 O'Connor and Jones, p. 17
8 Gerritsen, 2008, p. 62
9 Niewojt, p. 3
10 Tonkin and Landon, p. 208
11 Gott, 2005, p. 1205
12 ibid., p. 1203
13 Gammage, 2011, p. 166

14 ibid., p. 185
15 Flannery, 2010, p. 100
16 ibid.
17 Gammage, 2002, p. 9
18 ibid., p. 18

Chapter 6: The Heavens, Language, and the Law

1 Edwards, p. 203
2 ibid., p. 215
3 Young, p. 309
4 ibid.
5 Macinnis, p. 41
6 Bird-Rose in Edwards, p. 264
7 Stanner quoted by Bird-Rose in Edwards, p. 266
8 ibid., p. 214
9 Gammage, 2011, p. 321
10 Keen, 2004, p. 244
11 Le Griffon, p. 98
12 ibid., p. 187
13 McConvell and Evans, 1997, p. 46
14 ibid., p. 47
15 ibid., p. 417
16 Gammage, 2011, p. 150
17 Flood, 1983, p. 15
18 Archer, pp. 12–52
19 Barta, 2008a, p. 534
20 Stanner, 1979, p. 30
21 Stanner quoted in Edwards, pp. 225–36
22 Sturt, 1849, vol. 1, p. 124
23 ibid., p. 141
24 ibid., p. 113
25 ibid., p. 155
26 Mitchell, 1839, vol. 1, pp. 10–11
27 ibid., p. 11
28 ibid., p. 83

29 ibid., vol. 2, p. 159
30 Cruse, p. 17
31 Blay, 2005, p. 22

Chapter 7: An Australian Agricultural Revolution

1 Davies, Waugh, and Lefroy, 2005b, pp. 13–15
2 Peisley, 2011, *Microceris Lanceolata* pamphlet, p. 3

Chapter 8: Accepting History and Creating the Future

1 Menzies, pp. 232, 405–406
2 Menzies, p. 232
3 McNiven and Russell, p. 113
4 Snyder, p. 396
5 Peisley, 2010b, pp. 27–39
6 Bellanta, p. 7
7 Flannery, 2010, p. 34

Index

Note: *Page numbers in bold indicate images or captions.*